MATHEMATICS RESEARCH DEVELOPMENTS

SIMULATED ANNEALING

INTRODUCTION, APPLICATIONS AND THEORY

MATHEMATICS RESEARCH DEVELOPMENTS

Additional books in this series can be found on Nova's website under the Series tab.

Additional e-books in this series can be found on Nova's website under the eBooks tab.

MATHEMATICS RESEARCH DEVELOPMENTS

SIMULATED ANNEALING

INTRODUCTION, APPLICATIONS AND THEORY

ALEX SCOLLEN

AND

THOMAS HARGRAVES

EDITORS

nova science publishers
New York

NOTICE TO THE READER

The Publisher has taken reasonable care in the preparation of this book, but makes no expressed or implied warranty of any kind and assumes no responsibility for any errors or omissions. No liability is assumed for incidental or consequential damages in connection with or arising out of information contained in this book. The Publisher shall not be liable for any special, consequential, or exemplary damages resulting, in whole or in part, from the readers' use of, or reliance upon, this material. Any parts of this book based on government reports are so indicated and copyright is claimed for those parts to the extent applicable to compilations of such works.

Independent verification should be sought for any data, advice or recommendations contained in this book. In addition, no responsibility is assumed by the publisher for any injury and/or damage to persons or property arising from any methods, products, instructions, ideas or otherwise contained in this publication.

This publication is designed to provide accurate and authoritative information with regard to the subject matter covered herein. It is sold with the clear understanding that the Publisher is not engaged in rendering legal or any other professional services. If legal or any other expert assistance is required, the services of a competent person should be sought. FROM A DECLARATION OF PARTICIPANTS JOINTLY ADOPTED BY A COMMITTEE OF THE AMERICAN BAR ASSOCIATION AND A COMMITTEE OF PUBLISHERS.

Additional color graphics may be available in the e-book version of this book.

Library of Congress Cataloging-in-Publication Data

ISBN: 978-1-53613-674-6

Published by Nova Science Publishers, Inc. † New York

CONTENTS

PREFACE

The opening chapter of this book aims to present and analyze the application of the simulated annealing algorithm in solving parameter optimization problems of various manufacturing processes. The main specifications of the studied optimization problems are analyzed, and the tuning of the simulated annealing algorithm-specific parameters in solving process optimization problems is also discussed. Genetic algorithm, particle swarm optimization, etc., are also considered in process parameter design problems. Next, the authors describe applications of the simulated annealing method to biological macromolecular systems. The first important example is the reconstruction of an inappropriate three-dimensional (3D) structure of a functional site in a protein, built based on X-ray crystallographic experiments. The authors determine that the simulated annealing scheme is a powerful method that is applicable for resolving the local minimum problem in various scientific fields related to biological macromolecular systems. Following this, the development of the simulated annealing based MPPT method is detailed and simulations evaluating the performance of the technique are highlighted. Special focus is also paid to the parameters of the simulated annealing method, as well as which of these have the most influence based on the PV system configuration. Continuing, this collection proposes the application of simulated annealing algorithm to detect, from a fixed and random set of temperatures of the radiant zones of oven, the best temperatures that must be assigned to the heating zones. An example of the optimization of the heating stage of the high-density polyethylene (HDPE) grade sheet is presented. Another chapter discusses how, to reconstruct a porous media from two-dimensional (2D) images to three-dimensional (3D) images, several parameters (such as the spatial correlation) must be preserved. Otherwise, the 2D grain size distribution is known to be different from the 3D grain size distribution. Thus, a simulated annealing method to correct the grain size distribution from 2D to 3D is proposed. In the next study, a maximization methodology to recover energy (considering installation feasibility) was proposed to allocate recovery machines within water networks by using simulated annealing techniques. The authors suggest that simulated annealing, in the proposed methodology, is a powerful water management tool that may aid in the search for sustainability in irrigation networks. In the book's conclusion, the performance of a variable Markov chain length method in unconstrained optimization problems has been studied experimentally. For this purpose, eighteen benchmark optimization functions with different dimensions were employed to compare performance and convergence behavior of this algorithm to the standard simulated annealing algorithm with different Markov chain lengths.

Chapter 1 - The optimization of industrial processes is one of the key issues in a contemporary, highly competitive and globalized manufacturing sector. Since modern manufacturing processes are controlled by many control parameters at the input and deliver multiple quality characteristics i.e., responses at the output, the process optimization problem refers to the design of process parameters that meet the criteria for multiple responses. Due to the very high complexity of this problem, in recent years a growing number of researchers have employed artificial intelligence techniques to optimize the complex, both conventional and emerging, manufacturing processes. Based on a literature review, this chapter aims to present and analyze the application of a simulated annealing algorithm in solving parameter optimization problems of various manufacturing processes. The main specifications of the studied optimization problems are analyzed. The tuning of the simulated annealing algorithm-specific parameters, in solving process optimization problems, is also discussed in order to address issues of the algorithm convergence, robustness and effectiveness in reaching the actual global optimum. The benchmark of the performance of the simulated annealing algorithm with other evolutionary algorithms, such as genetic algorithm, particle swarm optimization, etc., in process parameter design problems is considered as well.

Chapter 2 – The authors describe applications of the simulated annealing (SA) method to biological macromolecular systems. The first example is the reconstruction of an inappropriate three-dimensional (3D) structure of a functional site in a protein, built based on X-ray crystallographic experiments. The SA method significantly extended the sampling space of the conformational space, and thus enabled the identification of the energetically-optimal 3D-structure of the functional site in the enzyme structure. The obtained structure facilitated various biophysical analyses, thereby leading us to theoretically resolve a discrepancy found in the biochemical experimental data.

The authors also briefly introduce another example of the use of the SA method to resolve structural issues, by constituting an extremely advanced energy function (Hamiltonian) for computer simulations of biological macromolecular systems. The SA method enhanced the explorations of the optimal parameter set defined in such an energy function, so its values could be fit with those obtained by the exclusively high level *ab initio* quantum mechanics calculations. This type of energy function (*i.e.*, an effective function) is very important to precisely describe molecular interactions (e.g., stacking interactions) that are frequently found and critical in the drug design field.

In an application to bioinformatics field, the SA scheme was combined with the Multivariate Curve Resolution-Alternating Least Squares (MCR-ALS) method, for the first time (our scheme is referred to as the SA-MCR-ALS method). The conventional MCR-ALS has been widely employed to analyze various multivariate data, such as spectrum data from circular dichroism (CD) and nuclear magnetic resonance (NMR) analyses, and image data from bioinformatics fields. However, it still fails to find the appropriate solutions in some cases. The authors describe an example where the MCR-ALS method fails, and how the SA-MCR-ALS method overcomes those difficulties.

In this manner, the SA scheme is a fundamental and powerful method that is intensively applicable to resolve the local minimum problem in various scientific fields related to biological macromolecular systems. In this chapter, the authors demonstrate such examples together with introductions of the relevant various technologies employed with the SA method.

Chapter 3 - Photovoltaic (PV) Systems are becoming increasingly common, particularly at residential scale, as society becomes more conscious of the environmental impact of conventional electricity generation. PV systems are made up of a number of PV cells joined together in series and parallel arrangements, which are then connected to a local load or utility grid through power electronics interfaces such as DC-DC converters and DC-AC inverters. These power electronics interfaces are necessary to enable the PV system to be controlled and to mitigate the impact of non-linear power-voltage (P-V) characteristics that arise in PV systems. In particular, in a residential environment, a PV system may be partially shaded by a tree or other structure which leads to a multi-peak P-V characteristic. Maximum Power Point Tracking (MPPT) is usually implemented in the DC-DC converter to ensure that the maximum power can be extracted from a PV system under these conditions. The area of MPPT is quite broad and many techniques have been adopted ranging from conventional techniques which can track a MPP under uniform conditions, to those that are designed specifically for the multi-peak scenario. The global searching capability of the Simulated Annealing (SA) technique makes it a good candidate method for solving the PV system MPPT problem in residential environments even with quite arbitrarily selected parameters. This chapter will detail the development of the SA based MPPT method and highlight simulations evaluating the performance of the technique. Special treatment will be given to the parameters of the SA method and which of these have most influence based on the PV system configuration.

Chapter 4 - The quality of the molded product in thermoforming process depends on the distribution of intercepted infrared radiation by the thermoplastic sheet surface in oven, the type of polymeric sheet, the optical properties and the temperature of the radiant zones and the mold temperature. To ensure that the energy flux intercepted by the different zones of the polymer is minimal, the authors propose the application of simulated annealing algorithm to detect, from a fixed and random set of temperatures of the radiant zones of oven, the best temperatures that must be assigned to the heating zones. For numerical heating analysis, the nonlinear heat conduction problem is solved by a specific 3D volumetric enthalpy-based finite element method. An example of optimization of the heating stage of the high-density polyethylene (HDPE) grade sheet is presented.

Chapter 5 - This chapter presents three Simulated Annealing applications, as well the Simulated Annealing theory. To reconstruct a porous media from two dimensional (2D) images to three-dimensional (3D) images, several parameters such as the spatial correlation must be preserved. Otherwise, the 2D grain size distribution is known to be different from the 3D grain size distribution. Therefore, it is exposed a Simulated Annealing method to correct the grain size distribution from 2D to 3D. Likewise, it is presented two porous media reconstruction methods using the Simulated Annealing optimization method in three-dimension (3D). Measuring geometrical and topological parameters in (2D), it is possible to generate a 3D microstructure model. The methods use different SA movements: one is based on pixels (for 2D) or voxels movements (for 3D) (PSA) and the other moves objects: circles (2D) or spheres (3D) (OSA). Both of them preserve the spatial correlation function and the PSA method also preserves the linear chord distribution and complementary distance distribution parameters.

Chapter 6 - Increase of sustainability in any system must be considered by all managers of different infraestructures as a primary objective. This concept consideration inside of water systems is also necessary. Therefore, water managers have to look for news strategies in order to improve the hydraulic and energy efficiency of their water networks. In this sense, changing pressure reduction valves by pumps working as turbines has been studied by researches since 1990s. However, the study of this recovery machines also needs strategies to help water managers to allocate them, maximizing the recovered energy. In this research, a maximization methodology to recover energy, considering installation feasibility, was proposed to allocate recovery manchines within water networks, by using simulated annealing techniques Combination of Epanet (hydraulic code) and simulated annealing in the proposed methodology is a powerful water management tool towards the search of the sustainability in irrigation networks. Particularly, by usingthis methodology 186 MWh/year could be thoretically recovered, reaching payback below 9 years. This energy generation could contribuite to develop more sustainable water systems, taking adavantageof optimization techniques.

Chapter 7 - The Simulated Annealing (SA) method is a heuristic optimization method inspired from the physical process by which a crystal is formed during slow cooling of melt until the global minimum energy state is reached. The most important feature of SA is perhaps the uphill movement capability using the Metropolis procedure. Although SA is generally known as a reliable method in finding the global optimum in both continuous and discrete functions and can produce extremely competitive solution as compared to other heuristic optimization methods, accurate tuning of its parameters to achieve a reasonable performance is crucial. One of these parameters is the Markov chain length (MCL), i.e., the number of iterations at a constant temperature for the system to reach a steady-state condition. In fact, the MCL should be long enough to satisfy a kind of equilibrium condition between the function value and temperature. In the most available SA algorithms, the MCL is determined experimentally and considered constant throughout the simulation. A strategy has previously been proposed that the MCL vary actively according to the optimization function variation. In this regard, temperature decrease is related directly to the function value variation history. One possible relation between these two in a minimization problem is the assumption of a constant temperature in trial solutions with decreasing or constant cost function values and changing temperature in trial solutions with increasing cost function. Because the temperature in this algorithm is changing variably with the objective function, an accelerated approach to the global optimum is expected due to the faster temperature decrease at high temperatures.

In this chapter, performance of this variable MCL method in unconstrained optimization problems has been studied experimentally. For this purpose, eighteen benchmark optimization functions with different dimensions were employed to compare performance and convergence behavior of this algorithm to the standard SA algorithm with different Markov chain lengths. Three different criteria were used for this comparison. The results show that the variable MCL algorithm can give rise to an improvement in convergence speed.

In: Simulated Annealing
Editors: Alex Scollen et al.
ISBN: 978-1-53613-674-6

Chapter 1

APPLICATION OF SIMULATED ANNEALING IN PROCESS OPTIMIZATION: A REVIEW

Tatjana V. Sibalija*
Belgrade Metropolitan University, Belgrade, Serbia

ABSTRACT

The optimization of industrial processes is one of the key issues in a contemporary, highly competitive and globalized manufacturing sector. Since modern manufacturing processes are controlled by many control parameters at the input and deliver multiple quality characteristics i.e., responses at the output, the process optimization problem refers to the design of process parameters that meet the criteria for multiple responses. Due to the very high complexity of this problem, in recent years a growing number of researchers have employed artificial intelligence techniques to optimize the complex, both conventional and emerging, manufacturing processes. Based on a literature review, this chapter aims to present and analyze the application of a simulated annealing algorithm in solving parameter optimization problems of various manufacturing processes. The main specifications of the studied optimization problems are analyzed. The tuning of the simulated annealing algorithm-specific parameters, in solving

* Corresponding Author Email: tsibalija@gmail.com; tatjana.sibalija@metropolitan.ac.rs.

process optimization problems, is also discussed in order to address issues of the algorithm convergence, robustness and effectiveness in reaching the actual global optimum. The benchmark of the performance of the simulated annealing algorithm with other evolutionary algorithms, such as genetic algorithm, particle swarm optimization, etc., in process parameter design problems is considered as well.

Keywords: simulated annealing, process optimization, multiple responses, process parameter design

Nomenclature

ABC	Artificial bee colony algorithm
ACO	Ant colony optimization
AI	Artificial intelligence
ANN	Artificial neural network
COA	Cuckoo optimization algorithm
GA	Genetic algorithm
GRA	Grey relational analysis
HH	Hoopoe heuristic
HS	Harmony search
PSO	Particle swarm optimization
RSM	Response surface methodology
SA	Simulated annealing
SFL	Shuffled frog leaping
SQP	Sequential quadratic programming
TLBO	Teaching–learning-based optimization
TS	Tabu search

1. Introduction

The optimization of the manufacturing processes is an important topic in a modern industrial sector, referring to determination of the optimal values of process parameters that meet the desired values for the response characteristics observed at the process output. Since the complexity of the manufacturing process grows, including a large number of process control parameters, possible noise factors and a large number of process responses, advanced methods are needed to address the optimization problem. The complexity of such problems increases with an increase in the number of responses, especially due to dependences among the responses. In the case of a process with multiple responses, multiresponse goals often conflict with each other. This is frequently solved by the engineers' judgment, previous experiences or various approximations. Such approaches cannot assure the desired outputs and cannot efficiently support the company in achieving a stable position in a highly competitive, globalized industrial market where consumer driven quality is one of the most emphasized topics. Therefore, advanced optimization approaches are needed to translate consumer expectations in terms of product quality characteristics into the design of manufacturing process parameters that will generate the expected quality. This especially applies for new, emerging manufacturing processes whose behavior has not been widely studied and analytical models of input-output relationships have not yet been established.

The appropriate design of process parameters is crucial for the effectiveness and quality of manufacturing in terms of producing the parts with the desired quality characteristics. Besides, the tuning of manufacturing process parameters has a significant effect on other parameters of the manufacturing system, such as productivity, overall equipment effectiveness, cost, lead time, time to market, etc., and finally, on the competitiveness of a company in the market.

Due to a complexity of mathematical models in the case of conventional manufacturing processes, and uncertainty or obscurity of process models in the case of emerging processes, artificial intelligence (AI) techniques have been increasingly used for process modeling and optimization. Many studies reported the need to engage evolutionary techniques such as genetic algorithm (GA), simulated annealing (SA), particle swarm optimization (PSO), etc. for process parameters design, to deal with highly nonlinear, multidimensional and ill-behaved complex optimization problems.

This chapter is dedicated to the methods for SA application in optimizing process parameters, for both traditional and modern manufacturing processes. After the introduction, the second section depicts basics of process parameters design, which is followed by SA algorithm background, procedure and algorithm-specific parameters presented in the third section. The fourth section gives a detailed and structured review of SA application in solving process optimization problems in the last decade. A discussion on the setting of SA algorithm-specific parameters in order to assure a fast convergence to the global optimum, and comparison with the other metaheuristics effectiveness in process optimization tasks are presented in the fifth section. Final remarks are given in the concluding section.

2. Process Optimization

In engineering, optimization implies the selection of a controllable input factors setting from a number of alternatives to achieve the desired outputs subjected to certain constraints. Process optimization refers to the determination of the setting of controllable process parameters that meet the specifications for the quality characteristics of products i.e., the desired process responses, and reduced variations. Modern manufacturing processes are typically controlled by multiple control factors i.e., process parameters ($x_1, x_2, ..., x_k$), they are usually affected by uncontrollable i.e., noise factors ($N_1, N_2, ..., N_q$), and at the output, multiple quality characteristics i.e., process responses ($y_1, y_2, ..., y_p$), are observed.

In order to assess the effect of process control factors on the responses in a structured and systematic manner, the designed experimentation is performed to collect the adequate set of input-output data used to establish the relationship between process responses and process parameters. The most frequently used methods for the experimental design and analysis are orthogonal arrays typically used within Taguchi robust parameter design methodology, and response surface methodology (RSM). Alternatively, if it is too expensive or impossible to conduct an actual experiment, in certain cases process historical data could be used to perform process optimization.

Today, since a majority of manufacturing processes deliver multiple responses, it is required to find a setting of process parameters that simultaneously produces the desired values for all responses, but also minimizes response dispersions i.e., reduce process variations. Typical causes of these variations are noise factors, so the optimal process parameters setting should be selected in such a way to minimize the effect of noise factors. This is effectively resolved by Taguchi robust parameter design, but only for a single response; for multiple responses, trade-offs are needed. The other traditional optimization methods do not directly address this issue. In the experimental data analysis, typically, each response is separately modeled as a function of the process parameters. But, since process responses are almost always correlated, the process parameters settings that optimize one response will degrade other responses. Optimizing the process purely with respect to any single response will lead to non-optimum solutions for the other responses. Therefore, the goal of multiresponse process optimization is to find the process parameters settings that achieve an optimal compromise of the response variables, so that each response is as close as possible to its ideal value with minimum dispersion.

The methodologies for process optimization evolved during the last two decades. Today, the optimization methods used for manufacturing processes optimization could be divided in two groups: conventional optimization approaches based on statistical methods, and non-conventional approaches based on artificial intelligence techniques. This detailed classification of the optimization methods used for the process parameter design of industrial processes in a broader sense, could be outlined as follows (Sibalija and Majstorovic, 2016):

1. Conventional, statistical or mathematical search- based methods:
   1.1. Iterative mathematical search techniques, i.e., goal programming;
   1.2. Experimental design (Design of Experiment – DoE) methods:
      - Methods based on response surface methodology (RSM),
      - Methods based on factorial design,

- Methods based on Taguchi methodology.

2. Non-conventional, artificial intelligence (AI) - based methods:
   2.1. Methods based on fuzzy logic;
   2.2. Methods based on artificial neural networks (ANNs);
   2.3. Methods based on metaheuristic search algorithms:
      - Genetic algorithm (GA),
      - Simulated annealing (SA),
      - Particle swarm optimization (PSO),
      - Ant colony optimization (ACO),
      - Tabu search (TS),
      - Recently developed evolutionary algorithms such as artificial bee colony algorithm (ABC), teaching–learning-based optimization (TLBO), etc.;
   2.4. Methods based on expert system (ES).

The process optimization approaches based on metaheuristic search algorithms, and, in particular based on SA algorithms, will be discussed in the following sections. In such approaches, based on experimental data, a process model is typically developed by ANNs, and then evolutionary algorithms are employed to find the optimal process parameters searching through a continual multidimensional space of solutions. Usage of metaheuristics in solving process optimization problems gained popularity in the last decade mainly due to their ability to effectively deal with continuous values of process parameters, in contrast to the majority of statistical approaches, and without referring to a particular mathematical model.

3. Simulated Annealing Algorithm

A simulated annealing belongs to the group of metaheuristic search techniques, as part of the artificial intelligence (AI) field. The metaheuristic algorithm is an iterative process that guides and modifies the operation of subordinate heuristics to efficiently produce high-quality solutions. At each iteration, the algorithm works with a complete or incomplete single solution or with a set of solutions at each iteration (Voß, 2001). Most of the metaheuristic techniques are based on nature or animal behavior, such as a genetic algorithm (GA), particle swarm optimization (PSO), ant colony optimization (ACO), artificial bee colony (ABC), etc. It has been proved that heuristic-based search algorithms are very useful in cases where conventional i.e., statistical optimization techniques are not effective, such as problems with high-dimensional search space with many local optima. Among them, SA algorithms have been frequently used to solve multiobjective optimization tasks in a wide variety of engineering problems.

The SA algorithm mimics the solids annealing procedure. It starts with selection of an initial random point at an initial (high) temperature and, according to the specified function, moves to a randomly selected new point. The difference between the function values of these two points ΔE is computed; if the objective function of a new point is superior to the current one, a new points is accepted, but, based on specified probability function, the algorithm can also accept the inferior point in order to extend the search area and avoid being trapped in a local optima in a multimodal response function. Once a new point is accepted, the temperature is reduced according to the cooling rate i.e., annealing schedule. As the temperature is reduced, the probability to move to an inferior point is also reduced thus controlling the randomness in a search, which mimics a slow cooling in a metal annealing process aiming to achieve a perfect crystalline state. Typically, to imitate this equilibrium, a number of points (i.e., algorithm iterations) are tested at a particular temperature before decreasing the temperature. At the end of a search when the temperature is nearly zero, the probability of accepting inferior points is very low, so the algorithm converges to an optimal solution (Spall 2003).

The most commonly used procedure of the SA algorithm could be outlined as follows:

1) Start: Tune the algorithm-specific parameters, such as the initial point, the termination criterion (e.g., the final temperature, the total number of algorithm iterations, the change in objective function),

the annealing schedule i.e., the annealing function and the temperature function, the number of iterations (n) to be performed at a particular temperature, and the reannealing interval.

2) Initial point: Calculate the objective function $f(x)$ of initial point.
3) New point: Calculate a new neighborhood point determined by a probability distribution proportional to the current temperature i.e., the annealing function; calculate the objective function $f(x)$ of a new point.
4) ΔE: Calculate the difference in the objective functions between the new and the current point ΔE i.e., $\Delta f(x)$.
5) Replacement: If $\Delta E < 0$, accept the new point, and go to step 6. If $\Delta E \geq 0$, create a random number r in the range (0, 1) and check whether $r \leq \exp(-\frac{E}{K \cdot T})$. If yes, go to step 6. If no, begin with a new point and go to step 3.
6) Iterations at a particular temperature: If the number of iterations is lower than the specified number of iterations (n) to be performed at a particular temperature, go to step 7. If not, perform iterations.
7) Temperature decrease: Periodically decrease the temperature according to the specified temperature update function. If the current temperature is higher than the final, go to step 3. If not, go to step 9.
8) Reannealing: In order to avoid local optima, raise the temperature in each dimension, i.e., perform reannealing after a number of points, specified by a reannealing interval parameter.
9) Termination: If the termination criterion is met (e.g., if the current temperature is lower than or equal to the specified final temperature, and/or the number of algorithm iterations is reached, and/or the change in the objective function is less than the specified value), terminate the procedure.

From the above procedure it is evident that the most important algorithm-specific parameters are the initial temperature, the final temperature or other termination criteria, the annealing schedule (the annealing function and the temperature update function), the number of iterations performed at each temperature, and the reannealing interval. In selecting the values for these algorithm-specific parameters, the following remarks should be considered (Sibalija and Majstorovic, 2016):

- The initial temperature must be high enough to allow a move to almost any neighborhood state. It determines the probability of accepting an inferior solution, preventing the algorithm from being trapped in a local optima, and forcing it to perform a wider search. But, if the initial temperature is too high, a search can move to any neighbor and become a random search.
- The final temperature is usually set to zero, but this can make the algorithm run for a very long time. Therefore, in practice the algorithm is terminated when a sufficiently low temperature is obtained, or a minimal change in the objective function value is achieved, or when the algorithm reaches a specified number of iterations.
- The temperature update function controls the annealing schedule in terms of a temperature decrement. The slower decrement rate is favorable for convergence to the global optimum, but it increases the algorithm run time.
- The number of iterations at each temperature should be large enough so that the system stabilizes at that temperature. Since this number is exponential to the problem size, sometimes it is recommended to perform only one iteration at each temperature and decrease the temperature very slowly. Alternatively, the number of iterations could be dynamically changed, where at the lower temperature a large number of iterations is performed so that the local optimum can be fully explored.
- The annealing function defines the generation of a new point in the next iteration. The algorithm selects the distance of the new random point from the current one by a probability distribution with a scale depending on the current temperature, which is controlled by the annealing function.

- The reannealing interval specifies the number of accepted points after which the temperature starts to increase. Reannealing sets the annealing parameters to lower values than the iteration number, thus raising the temperature in each dimension in order to avoid getting caught at a local optima. This parameter has to be considered in combination with the initial temperature. In general, performing reannealing too soon may not be favorable to identify an optimum, so a relatively high interval is recommended.

As described above, if the algorithm parameters are appropriately tuned, the probability of being trapped in a local optima is significantly reduced. It has been demonstrated that with an infinitely slow cooling, the algorithm is almost certain to find the global optimum. By combining the exploration features such as the random search and exploitation features like hill climbing, SA usually gives very good results in practice.

The major advantages of the SA algorithm could be summarized as follows (Spall, 2003; Mukherjee and Ray, 2006; Zandieh et al., 2009; Baseri, 2012):

- it can deal with arbitrary systems and objective functions, in terms of highly nonlinear and multimodal models, chaotic and noisy data and many constraints;
- it guarantees finding an optimal solution for a sufficiently large number of iterations;
- it can efficiently find the global optimum when the objective function has surrounding constraints;
- it is a generalized optimization technique, since it does not rely on any restrictive properties of the model, and no gradient calculation is required to determine its search direction;
- it is relatively easy to implement, and it is considered to be less sensitive to the size of a problem;
- it provides reasonably good solutions for the vast majority of optimization problems;

and the major shortcoming are:

- it is not a population-based algorithm, so it might not build up an overall view of the search space;
- convergence of the algorithm may be strongly affected by the algorithm-specific parameters, and no universally acceptable levels of parameters exist for different types of optimization problems;
- depending on the cooling schedule, it could be slow especially if the objective function is complicated;
- repeatability of the optimal solution obtained by SA algorithm with same initial conditions is not guaranteed.

4. Application of Simulated Annealing in Process Optimization

Simulated annealing, along with the other evolutionary techniques, gained significant attention in the last decade in solving various multicriteria problems in manufacturing and industrial engineering. Compared to the other global optimization algorithms, it is perceived that SA is easier to implement and less prone to be trapped in local optima, so it usually provides a good solution for many combinatorial problems (Yusup et al., 2012). Some of the tasks that have been successfully solved by SA implementation are manufacturing systems design, production planning, operation sequencing, job shop scheduling, etc. Metaheuristics have also been frequently used for the design of manufacturing process parameters in terms of finding the optimal values of process parameters that give the desired values of the responses observed at the output. This section presents a review of the SA application in manufacturing processes optimization in the last decade.

The review of studies that deal with SA implementation for process parameter design is performed according to the complexity and the type of problem from the optimization point of view. Studies are grouped into two major groups, according to the complexity of the optimization task:

optimization of processes where only one output characteristic i.e., response is considered, and optimization of processes with multiple output responses. Further division is done according to the development of a process model that presents the objective function for SA, as follows: (i) Type 1 – the analytical model of a process is known; (ii) Type 2 – the process model is developed using regression techniques; (iii) Type 3 – the process model is developed using artificial neural networks (ANNs); (iv) Type 4 – the process model is typically developed using ANNs, and the overall approach is problem independent, i.e., the optimization method could be used for a variety of process optimization problems.

4.1. Singe Response Optimization Problems

4.1.1. Single Response Problems - Type 1: The Process Model Function is Known

The simplest case of SA application for process optimization is when the process model function is known, as presented in the next few studies.

For a known analytical model of a process i.e., objective function, Saravanan (Saravanan et al., 2003) used GA and SA to optimize machining parameters of turning cylindrical stock in the presence of a single response. SA and GA were applied to determine the optimal machining parameters (depth of cut, cutting speed and feed) for continuous profile machining with respect to the minimum production cost subject to a set of practical constraints: cutting force, power constraints and tool tip temperature. Results showed that SA performed slightly better than GA. Tuning of SA parameters was not explicitly analyzed.

In order to design process parameters of a turning process on cylindrical workpieces, Kolahan and Abachizadeh (2008) developed SA algorithm based on the known analytical model of the objective function. The objective of this study was to find the optimal values of the considered process parameters (cutting speed, feed rate and cutting depth) in order to minimize the machining cost. The results showed that the application of the optimal process parameters found by SA-based procedure resulted in a significant reduction of the total machining cost. SA parameters' setting has not been discussed in this study.

Rao and Pawar (2010) optimized multi-pass milling process using SA, ABC and PSO algorithms. Several process parameters were considered, such as the number of passes, depth of cut, cutting speed and feed, aiming to minimize the production time i.e., to maximize the production rate. Based on the known process model i.e., objective function, SA was used to find the optimal process parameters. Its performance was benchmarked with ABC and PSO: the results showed that ABC and PSO showed superior performance in terms of the convergence rate (SA algorithm requires more iterations for convergence) and the accuracy i.e., quality of the obtained solution. SA parameters used in this study were: initial temperature = 200, and decrement factor = 0.1.

Mellal and Williams (2016) compared the performance of two recently developed bioinspired algorithms; Cuckoo optimization algorithm (COA) and Hoopoe heuristic (HH) with other frequently used evolutionary algorithms such as GA, SA ABC, PSO, teaching–learning based optimization (TLBO), using six real examples from the literature: optimization of parameters of two conventional processes (drilling process, grinding process) and four advanced machining processes (abrasive jet machining, abrasive water jet machining, ultrasonic machining, and water jet machining). All six studies refer to single response problems, where objective functions were considered known. The settings of SA algorithm-specific parameters used in these studies are not presented. In terms of quality i.e., accuracy of the optimal solution, the performance of SA in comparison to other algorithms could be summarized as follow: (i) Application of SA in optimization of grinding, drilling, and water jet machining has not been reported in this paper, so no comparison could be made for these three studies. (ii) For abrasive jet machining (AJM) in brittle materials where the material removal rate (MRR) was observed as the response, TLBO resulted in the best MRR value, followed by COA and SA that found an almost identical MRR value, and by HH and GA that found the lowest MRR value. (iii) For abrasive jet machining (AJM) in ductile materials where MRR was the response, TLBO outperformed the other algorithms, followed by COA, HH and SA that found almost the same MRR value, and by GA that found the lowest MRR value. (iv) For the abrasive water jet machining (AWJM) study where the objective was to minimize surface roughness Ra, the best results were obtained by COA, ABC, integrated ANN-SA, HH and integrated GA-SA approaches, followed by SA, and then by GA that found the highest Ra value. (v) For ultrasonic machining where the objective was to maximize MRR, COA and TLBO performed the best, followed by PSO, ABC, shuffled frog leaping (SFL), modified harmony search (HS_M), HH, SA, and GA, respectively.

In multipass turning process optimization, Jabri et al. (2017) used a hybrid SA-GA approach to find the optimal set of cutting parameters (cutting speed, feed rate, depth of cut, and number of passes) that will minimize the unit production cost. The mathematical model of a process is considered known. In order to improve a local search, SA was efficiently integrated in GA to avoid being trapped in a local optimum. The recommended SA parameters were: initial temperature = 500, cooling rate = 0.70.

4.1.2. Single Response Problems - Type 2: The Process Model Function Is Developed Using Regression Technique

As seen in the next few papers, regression modeling was frequently employed to generate the relationship between output response and process parameters. Then, the regression model presents an input for SA that searches for the optimal values of process parameters that meets the desired response value.

The abrasive water jet (AWJ) process has been optimized by Kolahan and Khajavi (2009) based on statistical methods i.e., Taguchi method and regression modeling to establish the relationship between process parameters (nozzle diameter, jet traverse rate, jet pressure and abrasive flow rate) and the process response i.e., depth of cut, in cutting 6063-T6 aluminum alloy. Then, SA was used to find the optimal process parameters setting that can produce a required depth of cut. The effectiveness of the proposed approach was confirmed, since the deviation from the desired output value was less that 0.5%. The settings of SA parameters in this study are as follows: initial temperature = 250, cooling rate = 0.98, and the termination criterion = 500 iterations.

Zain (Zain et al., 2010), performed the optimization of three parameters of the end milling process (cutting speed, feed and radial rake angle) in order to minimize surface roughness. By using the regression technique, the process model has been developed, presenting an input for SA that found the optimal values of process parameters. This resulted in a significant reduction of surface roughness, since the minimum surface roughness obtained by SA was much lower than the experimental sample data, regression modeling and RSM technique. The setting of algorithm-specific parameters has not been discussed.

Zain (Zain et al., 2011a) combined regression modeling, GA and SA to find the optimal process parameters (cutting speed, feed rate, and radial rake angle) of the end milling process aiming to minimize surface roughness Ra. The integrated SA-GA approach was successful in an estimation of the optimal process parameter conditions that lead to the minimum Ra value, within a relatively small computing time i.e., number of iterations. The best solution obtained by regression modeling is adopted as an initial solution in SA, and the setting of SA's own parameters were: annealing function = Boltzmann annealing, Reannealing interval = 100, temperature update function = exponential temperature, initial temperature = 100, and the acceptance probability function was SA acceptance.

Based on regression modeling, Zain (Zain et al., 2011b) used GA and SA to find the optimal values of process parameters of the abrasive water jet machining (traverse speed, waterjet pressure, standoff distance, abrasive grit size, abrasive flow rate), with an objective to minimize surface roughness Ra. Both GA and SA have been effective in estimating the minimum values of Ra compared to the experimental and regression results. In terms of computational time i.e., number of iterations, GA performed faster than SA, but SA performed better in terms of quality of the obtained solution. In other words, the minimum Ra value found by SA was much lower than the one found by GA. The setting of SA parameters was the same as in their others studies (Zain et al., 2011a, Zain et al., 2011c).

Aiming to optimize friction stir welding of dissimilar aluminum alloys, Padmanaban et al. (2014) used RSM to develop a regression model of a process, and SA to optimize the process parameters. Two control parameters were studied: tool rotation speeds and welding speeds, in order to improve the tensile strength of the joints. The optimum obtained by SA was successfully implemented. However, no details on the SA parameters have been presented.

In order to optimize TIG welding parameters, Joseph and Muthukumaran (2017) employed RSM and two evolutionary techniques GA and SA. Four process parameters were observed in Taguchi experimentations (current, voltage, welding speed, and gas flow rate), aiming to maximize the tensile strength of a weld. Based on experimental measurements, RSM was used to generate regression equations. Then, process parameters were optimized using GA and SA. SA outperformed GA in this study: SA obtained a better optimal solution that resulted in a higher tensile strength value, which was confirmed by the verification run. The selected values of SA parameters in this study were: initial temperature = 100, annealing function = Boltzmann annealing, reannealing interval = 100, temperature update function = exponential temperature update, and termination criterion refers to a small change in the function equation.

Ghalme et al. (2018) used the Taguchi method and SA to analyze wear behavior of silicon nitride. Two control factors of a process were observed (% volume of hBN, and load), with the aim to minimize wear rate. The Taguchi method was employed to design the experiment and to calculate a signal to noise ratio for the response. Based on this, a quadratic regression model is developed to model the process which presented an input for SA to find the optimal process parameter values. The settings of SA were not presented. The confirmation run verified the obtained results, and it was noted that SA found the optimal solution early – in 110th iteration.

4.1.3. Single Response Problems - Type 3: The Process Model Function Is Developed Using Anns

The next group of approaches implies usage of ANNs to develop a process model that presents the objective function for SA to find the optimal process parameters' setting.

Zain (Zain et al., 2011c) employed regression modeling, ANNs and SA to optimize the process parameters of AWJ (traverse speed, waterjet pressure, standoff distance, abrasive grit size and abrasive flow rate), in order to obtain the minimal surface roughness Ra. The results confirmed the effectiveness of ANN-SA approach in estimating minimal Ra values, within a relatively small number of iterations. In this study, the initial point of SA is set in a neighborhood of the potentially good solution obtained by using regression modeling. The SA parameters' setting was as follows: annealing function = Boltzmann annealing, reannealing interval = 100, temperature update function = exponential temperature, initial temperature = 100, acceptance probability function was SA acceptance.

Liao and Chen (2013) used a RSM and ANN&SA– based approach to determine the optimal process parameters of the micro-spot welding process. Four process parameters were studied (laser power, processing time, pulse frequency, and gas flow rate) in order to achieve the desired value of a shear strength. The proposed ANN&SA approach has better prediction and confirmation results than RSM method. The values of the algorithm-specific parameters used in this study were: annealing function = Boltzmann annealing, reannealing interval = 100, temperature updates function = exponential (temperature update decreases as $0.95^{iteration}$), initial temperature = 100.

In order to optimize the gas tungsten arc welding process, Sudhakaran et al. (2013) employed ANN and SA. Four process parameters were controlled (welding current, welding speed, gas flow rate and welding gun angle) aiming to predict the depth of penetration and obtained the desired value. ANNs were used to model the process, and SA was employed to find the optimal process parameter values. This approach is considered successful although it took 1300 iterations for SA to find the optimum. The following parameters were used in SA: initial temperature= 100, annealing function = fast annealing, reannealing interval = 100, temperature updates function = exponential temperature update (the temperature decreases as $0.95^{iteration}$).

Manoochehri and Kolahan (Manoochehri and Kolahan 2014) used ANN and SA to optimize the deep drawing process. The observed response was a minimum section thickness, and the study objective was to minimize thinning. The following process parameters were considered as control factors: blank holder force, punch radius, die radius, friction coefficient between punch–blank, die–blank, and holder–blank. Based on the results of finite element (FE) analysis, ANN has been developed to relate the process parameters to the observed response. Then, this neural model was integrated into the SA algorithm to optimize the process parameters. The results confirmed the effectiveness of the proposed approach, since the uniform wall thickness with minimum thinning was obtained using the optimal parameters setting. Besides, SA algorithm performed very fast since the optimal solution was obtained in 75th iteration. However, the setting of SA algorithm parameters was not presented.

Recently, Huang (Huang et al., 2015) designed an iteration approach based on ANN, GA and SA to optimize the thickness of blow molded bellows. The method employed ANN with the Bayesian regularization for process modeling. GA combined with the elitist strategy and SA (to improve a local search) was used to find the optimal values of seven process parameters that meet the required thickness. Hence, SA was used in this study for a local search. The selection of SA parameters was not presented.

4.1.4. Single Response Problems - Type 4: The Optimization Approach Is Problem Independent

In the previous three subsections the presented SA-based methods were designed to solve a particular optimization problem. Hence, they might not be suitable to address other optimization problems. In contrast to this, some researchers developed a generic, i.e., problem independent SA-based optimization method that could be applied to address different process optimization problems as follows.

In order to overcome the main shortcoming of RSM that is getting trapped in local minimum or maximum, Abbasi and Mahlooji (2012) used ANNs to improve the estimation in the RSM, and SA to maximize the estimated objective function. The effectiveness of the proposed approach was confirmed using three examples of different complexities. The tuning of SA parameters was not discussed.

4.2. Multiple Response Optimization Problems

4.2.1. Multiple Response Problems - Type 1: The Process Model Function Is Known

The next studies show the implementation of SA to solve process parameter design problems in case of multiple responses, for the known analytical model of a process.

In Corso's work (Corso et al., 2013), three optimization tools (i) sequential quadratic programming (SQP), (ii) GA, and (iii) SA were employed to find the optimal cutting parameters that minimize machining time while maintaining surface quality. The following process parameters were considered in this study: lateral increase, advance per revolution per tooth, and radius of the cut. The analytical models of the observed problem were considered as known. In this study, GA and SA performed better than SQP, because SQP found some of the local minima since it is a local method and depends upon gradient functions. Regarding the algorithm-specific parameters of SA, it was reported that the initial temperature was 1 and the temperature reduction factor was 0.95.

Pawar and Rao (2013) benchmarked the performance of TLBO with SA, GA and other metaheuristics in the parameter optimization of three machining processes. In terms of the quality of the obtained optimum, the results were as follows. (i) For abrasive water jet machining (AWJM) the objective was to maximize material removal rate (MRR) and five process parameters were observed (water jet pressure at the nozzle exit, diameter of abrasive water jet nozzle, feed rate of nozzle, mass flow rate of water, and mass flow rate of abrasives). TLBO found the highest MRR value, followed by SA, and GA showed a significantly lower result. (ii) For the grinding process, four control parameters were studied (wheel speed, work piece speed, depth of dressing doc and lead of dressing) aiming to maximize the production cost C_T and maximize the production rate in terms of workpiece removal parameter. PSO found the highest value for both responses followed by ABC; SA and TLBO showed slightly lower performance followed by harmony search (HS); GA showed the worst performance. In terms of speed of a convergence, TLBO requires only 30 iterations to converge to optimum while other algorithms required 60–80 iterations; SA algorithm reached the optimum in around the 60th iteration. (iii) For milling process, three control parameters were analyzed (feed per tooth, cutting speed and depth of cut) aiming to minimize the production time. Among many tested algorithms, the solution obtained using TLBO algorithm is slightly lower than solutions obtained by ABC, PSO, and SA algorithms; the other algorithms violated the constraints. In all three studies, various constraints were taken into account; SA parameters settings have not been presented.

4.2.2. Multiple Response Problems - Type 2: The Process Model Function Is Developed Using Regression Technique

In contrast to single response problems, regression modeling was used to develop the relationship between output response and process parameters only in a few cases, as follows.

Mukherjee and Ray (Mukherjee and Ray 2008) optimized parameters of the two-stage grinding process. In the stage I, nine process parameters and three responses were observed. In stage II, seven process parameters and five responses were analyzed. The process model was generated using multivariate linear regressions and desirability functions to incorporate desirability functions of various responses into a composite desirability function. Then, three metaheuristic search techniques, viz. GA, SA and a modified Tabu search (MTS) were employed to optimize the process parameters, and their performances were benchmarked. Results showed the following: MTS and SA showed a much higher probability to find the optimum in a limited number of iterations compared to GA; SA and MTS are easy to implement and require less algorithm parameters tuning than GA.

In terms of computational time, MTS was faster than SA which was much faster than GA. The following tuning of SA parameters was recommended: initial temperature = 45–55, final temperature = 40–50, cooling schedule function = power function, temperature increment = 0.1–0.2, termination criterion and number of iterations in a particular temperature = 2000 maximum number of main loop iterations and 50 iterations in a particular temperature.

In order to optimize the turning process, Kolahan et al. (2011) employed grey relational analysis (GRA) and SA. Aiming to maximize the material removal rate (MRR) and minimize surface roughness (Ra), three process parameters were observed (cutting speed, feed rate and depth of cut) in Taguchi experimentation. Based on the experimental results, GRA was applied to synthesize data from two responses into a grey relation grade that was used to develop the regression model. Then SA was successfully applied to find the optimal set of process parameters. The tuning of SA parameters was not depicted, but it was observed that SA converged to the optimum in the 50th iteration.

Somashekhar (Somashekhar et al., 2012) developed the SA- based approach to optimize parameters of micro-wire electric discharge process: gap voltage, capacitance, and feed rate. Using experimental data, statistical quadratic equations from regression analysis were developed for three responses: overcut, material removal rate (MMR), and surface roughness. The objectives were to maximize MRR and minimize surface roughness and overcut. A single objective function was generated from these regression models, and were successfully optimized using SA. The performance of SA was very good in terms of a fast convergence to the optimal solution as well as repeatability. SA algorithm was implemented with the following parameter values: initial temperature = 100, termination criterion = 10^{-3}, Boltzmann constant = 10, cooling rate = 0.95, number of cycles per temperature = 50.

Joseph and Muthukumaran (2016) compared performance of GA and SA in optimizing the pulsed current GTAW process. An experiment was conducted to assess the effect of four process parameters (peak current, gas flow rate, welding speed and base current) on two responses: tensile strength, and percentage of elongation of the joint. Using experimental results, regression models were developed for both responses. GA and SA were employed to find the optimal process parameters. Their performance was almost the same in terms of the quality of the optimal solution which was confirmed by the verification run. SA was faster than GA in terms of the number of iterations needed to reach the optimal solution. The tuning of SA parameters was as follows: initial temperature = 100, annealing function = fast annealing, reannealing interval = 100, temperature update function = exponential temperature update, and termination criterion refers to small change in function equation.

4.2.3. Multiple Response Problems - Type 3: The Process Model Function Is Developed Using Anns

The majority of multiresponse optimization problems addressed by the usage of SA refers to the implementation of ANNs to model the relationship between output responses and process control parameters. Such a neural process model presents an objective function for SA that searches for a global optimal values of process parameters.

Yen (Yen et al., 2006) designed neural networks (NNs) and SA-based methods to perform multiresponse optimization in injection molding. In the modeling stage, results from the finite element method (FEM) were used by abductive NNs to model relationships between the runner system parameters (diameter and length of the runner-system, gate diameter and length, and material thickness) and the warp performance of the molding process i.e., warpage of the molded parts. In terms of warpage, the following responses were observed: the warp, the temperature difference, and the total time. Then, in the optimization stage, NNs model presents an objective function for SA that finds the optimal process parameters values in order to minimize all three responses. The SA parameters setting has not been presented.

Yang (Yang et al., 2009) used ANN and SA to optimize dual-response electric discharge machining (EDM) process. Four process parameters of EDM were studied: discharge current, source voltage, pulse-on time and pulse-off time. The objectives were to maximize the material removal and to minimize the surface roughness. Using experimental data, a counter-propagation neural network was employed to model the system, and SA was used to find the optimal process parameter values that can satisfy the specifications for two responses. The obtained optimal values were successfully verified in a test run. The SA parameters were set as follows: initial temperature = 400, termination criterion = 10^{-5}, Boltzmann constant = 10, cooling rate = 0.9, number of cycles per temperature = 100.

A study on the optimization of friction welding was performed by Sathiya (Sathiya et al., 2009). Four process parameters were considered: heating pressure, heating time, upsetting pressure, upsetting time. The objectives were to maximize the tensile strength of the friction joint and to minimize the metal loss. ANNs were utilized to model the relationships between two responses and four process parameters. Then, three evolutionary computational techniques, namely, GA, SA and PSO were employed to find the optimal process parameter condition in order to meet the requirements for two responses. The results showed that GA outperformed the other two metaheuristic methods in terms of maximized tensile strength and minimized metal loss. The parameters used in SA were as follows: initial temperature = 1000, final temperature 1, and decaying ratio Cr = 0.95.

Chen (Chen et al., 2010) presented an ANN&SA-based method for the optimization of wire electrical discharge machining (WEDM). The following process parameters were analyzed: the pulse on time, the pulse off time, arc off time, the servo voltage, the wire feed rate, the wire tension and the water pressure. Three responses were observed: cutting velocity, surface roughness and roughness maximum. Based on Taguchi's experimental design, a process model was generated by ANNs. SA techniques were used to find the optimal process parameters using a neural process model. The confirmation run was performed using the optimal parameters condition, and it verified the predicted results. The annealing approach was as follows: annealing function = Boltzmann annealing, reannealing interval = 100, temperature updates function = exponential (temperature update decreases as $0.95^{\text{iteration}}$), initial temperature = 100.

Aiming to optimize the dressing conditions to improve the grinding performance, Baseri (2012) used ANN to develop the process model, and

SA to optimize the process parameters. The objective was to minimize the grinding force subject to surface roughness constraint. In this study the grinding parameters were kept fixed and not analyzed, but three dressing parameters were observed: dressing cross-feed rate, dressing speed ratio, and depth of dressing. The implementation of optimal parameters found by SA was successful. In this study the SA setting was as follows: initial point = [0.1 0.1 0.1], initial temperature = 100, Boltzmann constant = 1, stop temperature = $1e^{-8}$, cooling rate = 0.8, maximum number of tries within one temperature = 300.

In order to optimize the parameters of a wire electrical discharge machining (WEDM), Yang (Yang et al., 2012) used a method based on ANN and SA. Seven parameters were selected as process control factors: the pulse on time, the pulse off time, arc off time, the servo voltage, the wire feed rate, the wire tension, and the water pressure. The experiment was designed using a Taguchi orthogonal array, and three responses were measured: metal removal rate (MRR), roughness average, and corner deviation. The objectives were to maximize MRR and minimize the other two responses. Using experimental results, ANNs were developed to model the process responses vs. process parameters, which present an objective function for SA. The results obtained by the ANN&SA method were benchmarked to RSM, where ANN&SA outperformed RSM in terms of quality. The optimal solution obtained by SA was confirmed in a verification run. SA parameters were tuned as follows: annealing function = Boltzmann annealing, reannealing interval = 100, temperature updates function = exponential (temperature update decreases as $0.95^{iteration}$), initial temperature = 100.

Teimouri and Baseri (2013) used ANN, SA and PSO to predict the percentage of surface roughness variation in magnetic abrasive finishing, where four process parameters were studied: electromagnet's voltage, mesh number of abrasive particles, poles rotational speed, and weight percent of abrasive particles. Based on ANN models, SA and PSO were employed separately to find the optimal process parameters. Results showed that both algorithms were effective and accurate in finding the optimum. The setup parameters for SA implementation were: initial point = [0 0 0 0], initial temperature = 500, Boltzmann constant = 1, stop temperature = $10e^{-8}$, cooling rate = 0.9, number of tries within one temperature = 300.

Babajanzade Roshan (Babajanzade Roshan et al., 2013) optimized a friction stir welding process using SA-based method. The process parameters taken into account were: tool pin profile, tool rotary speed, welding speed, and welding axial force. The objective of the study was to obtain the desirable mechanical properties of aluminum 7075 plates, so three responses were observed: tensile strength, yield strength, and hardness of welded zone. Based on the experimental data, adaptive neuro-fuzzy inference systems were developed for each response separately to generate the process models that served as objective function for SA to find the optimal values of four process parameters. The obtained optimal solution was verified by confirmation tests. The adopted SA parameters' conditions were as follows: initial point = [0 0 0 0], initial temperature = 500, Boltzmann constant = 1, stop temperature = $10e^{-8}$, cooling rate = 0.9, maximum number of tries within one temperature = 300.

Coppel et al. (2016) presented an on-line adaptive control optimization system for micro-milling, which considers the real cutting-tool wear state in terms of parts quality and adapts the cutting conditions (feed rate, spindle speed) to reach a minimum production cost. The system consisted of (1) a monitoring sensor system, (2) an estimation module with ANN models, (3) an optimization module with evolutionary optimization algorithms, and (4) a CNC interface module. In an optimization module, different algorithms have been tested: GA, SA and PSO. In summary, the PSO algorithm showed better efficiency in terms of computing time than the GA and SA algorithms. The adopted SA parameters were: annealing function = fast annealing, temperature function = exponential, initial temperature = 300, reannealing interval = 50.

4.2.4. Multiple Response Problems - Type 4: The Optimization Approach Is Problem Independent

The above mentioned SA- based methods have been developed to optimize a particular process. Hence, they cannot be easily adopted to address other multiresponse optimization problems. The next paragraphs show SA-based multiresponse optimization approaches that are problem independent i.e., approaches that could be applied to solve a variety of multiresponse process parameter design problems.

Chang (Chang 2008) proposed a generic method to optimize dynamic processes i.e., signal-response systems, as defined by Taguchi. The approach involved signal-response system modeling using ANNs to develop the response model. Then, a performance measure was developed for each type of response according to the specification (smaller the better, larger the better, nominal the best) to evaluate the predicted responses. The third phase involves transforming. Finally, the performance measures were transformed into energy values in SA in order to obtain the optimal process parameters values that can meet the specifications for the observed responses. A numerical example of a dynamic vacuum testing unit was used to verify the applicability of the proposed approach: eleven process parameters were taken as control factors including parameters with qualitative and quantitative values, two noise factors were observed, and one signal factor. At the output, one response has been evaluated. Although numerical study implied only one response, this approach is also applicable for multiple responses. The SA parameters tuning has not been presented in this work.

In order to improve the methodology for multiresponse process optimization that was based on GA (Sibalija et al., 2011; Sibalija and Majstorovic, 2012a), the SA algorithm has been employed as an optimization tool instead of GA. The methodology consists of three stages: (i) Taguchi quality loss function, principal component analysis (PCA) and grey relational analysis (GRA) are used to adequately present, uncorrelate and synthesize the responses into a process performance measure; (ii) back propagation, feed forward ANNs were used to generate the process model i.e., relationship between the process performance measure and process parameters; (iii) metaheuristic optimization algorithms were used to find the optimal values of process parameters. Using four case studies (two studies from semiconductor industry that refer to microchips assembly wire bonding process; one study that implies drilling process performed by pulsed Nd: YAG laser on a nickel-based superalloy; and one that refers to the automatic enameling process in a cookware production), it has been demonstrated that

SA performed better than GA within this methodology (Sibalija and Majstorovic, 2012b). In particular, SA outperformed GA in terms of the quality of the optimal solution and robustness of the algorithm (significantly lower effect of SA algorithm's parameters' setting on the quality of a final solution than for GA). Besides, SA showed a slightly better speed of a convergence to the final solution than GA, and it needed less computational time. Regarding SA parameters' setting, several values for the major parameters have been tested, and results are as follows: initial temperature of 100 showed the best results, but the value of 500 also performed well; annealing function: Boltzmann annealing algorithm performed slightly better than the fast annealing algorithm; temperature function: both the Boltzmann and the fast temperature function performed very well; reannealing interval: interval of 10 in combination with initial temperature of 100 showed the best results. The initial point has been set in proximity to a potentially good solution.

In their next study, Sibalija (Sibalija et al., 2014) used the above SA-based methodology to optimize the laser shock peening process applied on nickel based superalloy Nimonic 263 sheets. Three process parameters were studied (voltage, focus position, and pulse duration) in order to reach the desired values for seven responses: circularity, grain size, roughness average, average maximum roughness valley depth, average maximum roughness peak height, skewness, and kurtosis. Relying on the results of the previous studies presented above, different settings of SA parameters have been tested: initial temperature = 100 and 500; annealing function = Boltzmann and the fast annealing function; temperature function = Boltzmann and the fast temperature function; reannealing interval = 10 and 100, in combination with the initial temperature values of 100 and 500; termination criteria = 5000 iterations or change in objective function less than 10^{-12}, whichever is earlier. The SA algorithm that showed the best results was set as follows: initial temperature = 100; annealing function = fast annealing; temperature function = Boltzmann temperature function; reannealing interval = 10.

5. Discussion

5.1. Setting of Simulated Annealing Algorithm Parameters

The SA algorithm has been widely used in different multivariate optimization problems, mainly due to its simplicity but also due to its effectiveness. As presented in the previous section, SA could be used to solve multivariate problems with both continuous and discrete variables; it can effectively solve non-linear optimization problems even in a case of a multi-modal response function due to its stochastic nature and guided probabilistic moves. Besides, it is convenient for a variety of applications since it doesn't require gradient calculation to determine the search direction. Moreover, it is perceived that SA is less sensitive to the problem size than the other metaheuristics.

Previous studies showed that SA performance could be strongly affected by the setting of the algorithm's own parameters and, in particular, by the state of an initial point due to the one-point nature of the search procedure (Spall, 2003; Mukherjee and Ray, 2006; Zandieh et al., 2009). That is why it is important to analyze these issues in terms of their effects on the algorithm convergence and the quality of the obtained optimal solution. In SA-based methods used for process optimization discussed in the previous section, only some studies presented algorithm settings.

Here is the summary of the SA algorithm parameter values used in single response optimization problems:

- Initial temperature. In the observed studies, the most frequently used value for initial temperature was 100, but values of 200, 250, and 500 have been also reported.
- Reannealing interval. The value of 100 was the most frequently used, typically in combination with an initial temperature value of 100.

- Annealing function. Boltzmann annealing function was reported in the majority of studies, except in one study where the fast annealing function was used.
- Temperature update function. The exponential temperature function was utilized in all presented studies. For cooling rate, values 0.70, 0.90, 0.95 and 0.98 were specified.
- The termination criteria. The specified number of iterations (e.g., 500, 1000, etc.) and the small change in the objective function are the most frequently used criteria.
- The setting of an initial point of the algorithm has not been commented in the observed single response studies.

For multiresponse problems, the following algorithm settings have been reported:

- Initial temperature. The value of 100 was commonly used, along with other values such as 45-55, 300, 400, 500, and 1000.
- Reannealing interval. The most frequently used interval was 100, mainly in combination with initial temperature of 100. Besides, a value of 50 in combination with an initial temperature of 300 was also reported, as well as the interval of 10 in combination with the initial temperature 100 and 500.
- Annealing function. In a majority of cases, the Boltzmann annealing function was used; three studies reported usage of the fast annealing function.
- Temperature update function. The exponential temperature function, where the temperature is reduced exponentially at each iteration, was utilized in a majority of cases (with cooling rate values of 0.95, 0.90 and 0.80), although in one study the Boltzmann function and the fast temperature function showed better results than the exponential temperature function.
- The termination criteria. The defined number of iterations (e.g., 2000, 5000, etc.) in combination with the small change in the objective function (e.g., 10^{-3}, 10^{-5}, 10^{-8}, 10^{-12}), and the specified

number of cycles per temperature (e.g., 50, 100, 300) are the most frequently used criteria.

- Initial point. The initial point setting was specified only in three studies (e.g., [0.1 0.1 0.1], [0 0 0 0]), however the reason for selecting the initial state has not been discussed.

The above analysis is divided into two type of problems, since multiresponse problems are expected to be more complex than single response problems. Hence, it was assumed that this difference in a problem's complexity could affect the tuning of SA algorithm. However, as it could be seen above, major differences in algorithm parameters' setting between single and multiple response problems haven't been found.

In the observed process optimization studies, the authors only reported the selected values for the algorithm parameters, but the rationale for their selection was not discussed.

An attempt in this direction has been done by Sibalija and Majstorovic (2016) in a structured manner, where different values of algorithm parameters have been tested, and results on six case studies have been summarized. The following values of SA parameters have been studied: initial temperature = 10, 100 and 500; temperature update function = exponential function, fast function, and Boltzmann function; annealing function = fast annealing function, and Boltzmann annealing function; reannealing interval: = 10 and 100. Therefore, 36 SA algorithms were developed for each multiresponse optimization problem. The results, in terms of the beneficial effect of the algorithm parameters on the quality of a final solution and on the convergence to the global optimum, could be summarized as follows: (i) Annealing function: Boltzmann annealing function performed slightly better that the fast annealing that also performed very well, so both functions could be recommended. (ii) Temperature function: The best results showed Boltzmann temperature followed by the fast temperature function; the exponential function performed slightly worse. Hence, Boltzmann and fast temperature functions could be suggested. (iii) Reannealing interval and initial temperature: The reannealing interval of 10 in combination with the initial temperature of 100 showed the best results, followed by the combination with the initial temperature of 500. The interval value of 100 with the initial temperature values of 100 and 500 also showed good results. As it could have been expected, the initial temperature of 10 gave significantly lower results in all combinations. It has to be highlighted that, under the above recommended algorithm settings, SA found an actual global optimum very efficiently i.e., the speed of convergence to the optimum was very high (the optimum was found in the first 150 iterations, for all six studies). Having in mind these findings and also the fact that SA parameter setting is still problem-dependent, the above recommended tuning of the algorithm-specific parameters could be taken as a potential guideline for this type of optimization problems.

5.2. Benchmark with Other Metaheuristic Search Algorithms

There are many evidences from the literature that metaheuristic search techniques have been successfully used instead of the statistical method for the process parameter design (Mukherjee and Ray, 2006; Rao and Kalyankar, 2014), mainly due to their ability to search in a multidimensional continual space of solutions whereas the statistical methods typically consider only the space of discrete variables (except for RSM). Among a variety of evolutionary algorithms, GA is the most frequently used for machining processes optimization, followed by SA; the other techniques such as PSO, ACO and ABC have been used since recently (Yusup et al., 2012; Chandrasekaran et al., 2010).

In a significant number of studies presented in the previous section, SA performance was compared to the performance of some other metaheuristic algorithms or to RSM, which is still one of the major statistical techniques for process parameters optimization. The comparison of SA with other algorithms, in terms of the quality of the obtained optimum and the speed of convergence i.e., the number of iterations needed to reach the optimum, are summarized in Table 1 and the following conclusions could be drawn.

- SA and GA: SA performed better than GA in a vast majority of cases, both in terms of the quality of the obtained solution and in terms of the speed of a convergence. From 21 studies where SA and GA were benchmarked, only in one study did GA perform better, and in two studies both algorithms found the same solutions.
- SA and RSM: SA found better solutions than RSM in all seven studies presented in this comparison. The convergence could not be compared since RSM in not a metaheuristic search but a statistical technique.
- SA and PSO: PSO outperformed SA in terms of the quality of the solution in five studies, and in one study they found the same optimum. The speed of a convergence for both algorithms was approximately equal.
- SA and ABC: ABC found better solutions in all five cases; both algorithms performed in a similar manner regarding the speed of convergence.
- SA and TLBO: In four case studies, TLBO found the better solution than SA; in one study both algorithms found the same solution, and in one study SA found the better solution. The convergence was analyzed in only one study where TLBO showed a faster convergence to the optimum than SA.
- SA and COA: In three cases COA found the better solution than SA, and in one study both algorithms found the same solution. The convergence issue has not been discussed in these four studies.
- SA and HH: The comparison of SA and HH in four studies showed even results in terms of quality of the obtained solution. The speed of convergence has not been discussed in these studies.
- SA and HS: In terms of the solution quality, in one study SA found a better solution, but in another one, the modified HS algorithm performed better. A similar speed of convergence was reported for both algorithm in one study.
- SA and other metaheuristic algorithms: In one study it was reported that SFL found a better solution than SA. SA outperformed SQP in terms of solution quality in one optimization problem. One study

benchmarked the performance of MTS and SA, where both algorithms converged to the same solution but SA convergence was slower than MTS.

Considering the application of SA and other metaheuristics in solving process optimization problems, it could be seen that SA outperformed GA in a majority of studies; its performance was equal to HH and HS, but it was inferior to PSO, ABC, TLB and COA. However, these findings should be taken conservatively due to the following reason. The setting of SA parameters has not been discussed in some of the above studies; in others it has been presented but without proving that the chosen setting is the best one for the observed problem. Since SA performance depends on the setting of SA's own parameters, if this setting is not appropriate the performances of SA, both in terms of the solution quality and the speed of a convergence, is jeopardized. Therefore, this issue deserves significant attention in future applications of SA for solving process optimization problems.

With reference to the above, an important issue that needs further attention is the comparison of the robustness of the metaheuristic algorithms. Beside the quality of a final solution and the speed of a convergence as two major criteria in comparing SA and GA, Sibalija and Majstorovic (2016) also discussed the robustness of SA and GA in terms of the dispersion of obtained optimal solutions caused by different settings of the algorithm-specific parameters. Based on results obtained from six case studies, it has been shown that SA demonstrated significantly better robustness than GA in four cases (i.e., all tested combinations of SA's parameters gave approximately the same results, in contrast to GA where variation of algorithm's parameters significantly affects the result quality), and slightly better robustness in two studies. The comparison of robustness of SA with other evolutionary algorithms has not been found in the above literature review that refers to metaheuristics application in solving process optimization problems. Therefore, future research should address this problem.

Table 1. Comparison of SA performance with other evolutionary methods used in optimization studies from the literature

Author and year	Problem type	Process	SA performance: quality of the obtained solution	SA performance: speed of a convergence
Saravanan et al., 2003	Single response - Type 1	turning cylindrical stock	• SA better than GA	-
Rao and Pawar, 2010	Single response - Type 1	multi-pass milling	• SA worse than ABC and PSO	• SA worse than ABC and PSO
Mellal and Williams, 2016	Single response - Type 1	abrasive jet machining in brittle materials	• SA worse than TLBO • SA and COA equal • SA better than HH and GA	-
	Single response - Type 1	abrasive jet machining in ductile materials	• SA worse than TLBO and COA • SA and HH equal • SA better than GA	-
	Single response - Type 1	abrasive water jet machining	• SA worse than COA, ABC, ANN-SA, HH, GA-SA • SA better than GA	-
	Single response - Type 1	ultrasonic machining	• SA worse than COA, TLB, PSO, ABC, SFL, HS_M • SA and HH equal • SA better than GA	-
Zain et al., 2010	Single response - Type 2	end milling	• SA better than RSM and regression modeling	-
Zain et al., 2011b	Single response - Type 2	abrasive water jet machining	• SA better than GA	• SA worse than GA
Joseph and Muthukumaran, 2017	Single response - Type 2	TIG welding	• SA better than GA	-
Liao and Chen, 2013	Single response - Type 3	micro-spot welding	• SA better than RSM	-
Corso et al., 2013	Multiple responses - Type 1	cutting processes	• SA and GA equal • SA better than SQP	-

Author and year	Problem type	Process	SA performance: quality of the obtained solution	SA performance: speed of a convergence
Pawar and Rao, 2013	Multiple responses - Type 1	abrasive water jet machining	• SA worse than TLBO • SA better than GA	-
	Multiple responses - Type 1	grinding	• SA worse than PSO and ABC • SA and TLBO equal • SA better than HS and GA	• SA worse than TLBO • SA equal to or better than PSO, ABC, HS and GA
	Multiple responses - Type 1	milling	• SA worse than PSO and ABC • SA better than TLBO	-
Mukherjee and Ray, 2008	Multiple responses - Type 2	two-stage grinding	• SA and MTS equal • SA better than GA	• SA worse than MTS • SA better than GA
Joseph and Muthukumaran, 2016	Multiple responses - Type 2	pulsed current GTAW	• SA and GA equal	• SA better than GA
Sathiya et al., 2009	Multiple responses - Type 3	friction welding	• SA worse than GA and PSO	-
Yang et al., 2012	Multiple responses - Type 3	wire electrical discharge machining	• SA better than RSM	-
Teimouri and Baseri, 2013	Multiple responses - Type 3	magnetic abrasive finishing	• SA and PSO equal	-
Coppel et al., 2016	Multiple responses - Type 3	micro-milling		• SA worse than PSO • SA and GA equal
Sibalija and Majstorovic, 2016	Multiple responses - Type 4	thermosonic copper wire bonding on copper pins	• SA and GA equal • SA better than RSM	• SA worse than GA
	Multiple responses - Type 4	thermosonic copper wire bonding on silver plated pins	• SA better than GA and RSM	• SA better than GA
	Multiple responses - Type 4	Nd:YAG laser drilling	• SA and GA equal • SA better than RSM	• SA better than GA
	Multiple responses - Type 4	Nd3+:YAG laser shock peening	• SA and GA equal	• SA better than GA
	Multiple responses - Type 4	thermosonic gold wire bonding process	• SA better than GA and RSM	• SA better than GA
	Multiple responses - Type 4	automatic enameling	• SA better than GA	• SA better than GA

Conclusion

This chapter presents the application of the SA algorithm in solving process optimization problems, considering both conventional and modern, advanced manufacturing processes. It gives an overview of the latest ten year research (from 2008 to 2018) on this topic, but a few older studies are reported as well. The literature review was divided into two main groups according to the nature of the optimization problem (single response, and multiple response optimization problems), which was further separated in four types of problems according to the objective function issue (Type 1 - objective function model is known; Type 2 - objective function generated using regression modeling; Type 3 - objective function generated using ANNs, and, Type 4 - problem-independent optimization approaches).

Since it is known that the effectiveness of the SA algorithm is influenced by the tuning of its own parameters, this problem was observed for each study. Based on these results, general recommendations for setting SA algorithm-specific parameters in optimizing parameters of manufacturing processes were drawn.

The effectiveness of SA in process parameter design was benchmarked with other metaheuristic search algorithms. It has been found that GA is the most widely used, followed by SA, and then by PSO, ACO, ABC, TLBO, etc. Based on the findings presented in studies from the literature review, it could be seen that SA outperformed GA in a majority of the observed problems, but it was inferior in comparison to more recent algorithms such as PSO, ACO, ABC, etc. However, the comparison of SA to these recently developed algorithms was presented in a relatively smaller number of studies (around five studies), so these findings should be confirmed or denied in future studies. Besides, further research is needed on the assessment of the algorithm's sensitivity to the algorithm-specific parameters tuning, and comparison between evolutionary algorithms in solving process optimization problems should include this aspect as well.

To summarize, in the vast majority of studies it has been demonstrated that the application of SA in optimizing manufacturing processes gave very good results, so SA is still considered to be a promising solution for these kinds of problems. Therefore, future application of SA and other evolutionary algorithms in process parameters design could be expected, especially in the optimization of complex, non-conventional, emerging manufacturing processes where the analytical model of a process i.e., objective function is unknown.

References

Abbasi B., Mahlooji H. (2012) Improving response surface methodology by using artificial neural network and simulated annealing. *Expert Systems with Applications* 39: 3461–3468. doi:10.1016/j.eswa.2011.09.036.

Babajanzade Roshan S., Behboodi Jooibari M., Teimouri R., Asgharzadeh-Ahmadi G., Falahati-Naghibi M., Sohrabpoor H. (2013) Optimization of friction stir welding process of AA7075 aluminum alloy to achieve desirable mechanical properties using ANFIS models and simulated annealing algorithm. *International Journal of Advanced Manufacturing Technology* 69(5-8): 1803-1818.

Baseri H. (2012) Simulated annealing based optimization of dressing conditions for increasing the grinding performance. *International Journal of Advanced Manufacturing Technology* 59:531–538. doi: 10.1007/s00170-011-3518-9.

Chandrasekaran M., Muralidhar M., Krishna C. M., Dixit U. S. (2010) Application of soft computing techniques in machining performance prediction and optimization: a literature review. *International Journal of Advanced Manufacturing Technology* 46: 445–464.

Chang H. H. (2008) A data mining approach to dynamic multiple responses in Taguchi experimental design. *Expert Systems with Applications* 35(3):1095–1103. doi: 10.1016/j.eswa.2007.08.005.

Chen H. C., Lin J. C., Yang Y. K., Tsai C. H. (2010) Optimization of wire electrical discharge machining for pure tungsten using a neural network integrated simulated annealing approach. *Expert Systems with Applications* 37(10): 7147–7153.

Coppel R., Abellan-Nebot J., Siller H., Rodriguez C., Guedea F. (2016) Adaptive control optimization in micro-milling of hardened steels—evaluation of optimization approaches. *International Journal of Advanced Manufacturing Technology* 84:2219–2238. doi: 10.1007/s00170-015-7807-6.

Corso L. L., Zeilmann R. P., Nicola G. L., Missell F.P., Gomes H.M. (2013) Using optimization procedures to minimize machining time while maintaining surface quality. *International Journal of Advanced Manufacturing Technology* 65:1659–1667. doi: 10.1007/s00170-012-4288-8.

Ghalme S., Mankar A., Bhalerao Y. (2018) Integrated Taguchi-simulated annealing (SA) approach for analyzing wear behaviour of silicon nitride. *Journal of Applied Research and Technology* 15: 624–632.

Huang H. X., Li J. C., Xiao C. L. (2015). A proposed iteration optimization approach integrating backpropagation neural network with genetic algorithm. *Expert Systems with Applications,* 42, 146–155.

Jabri A., Barkany A. E., Khalfi A. E. (2017) Multipass Turning Operation Process Optimization Using Hybrid Genetic Simulated Annealing Algorithm. *Modelling and Simulation in Engineering,* vol. 2017, Article ID 1940635, 10 pages, doi:10.1155/2017/1940635.

Joseph J., S. Muthukumaran (2016) Optimization of pulsed current GTAW process parameters for sintered hot forged AISI 4135 P/M steel welds by simulated annealing and genetic algorithm. *Journal of Mechanical Science and Technology* 30 (1):145~155. doi: 10.1007/s12206-015-1218-3.

Joseph J., S. Muthukumaran (2017) Optimization of activated TIG welding parameters for improving weld joint strength of AISI 4135 PM steel by genetic algorithm and simulated annealing. *International Journal of Advanced Manufacturing Technology* 93:23–34. doi: 10.1007/s00170-015-7599-8.

Kolahan F., Abachizadeh M. (2008). Optimizing turning parameters for cylindrical parts using simulated annealing method. *World Academy of Science, Engineering and Technology,* 46: 437–439.

Kolahan F., Khajavi A. H., (2009). A statistical approach for predicting and optimizing depth of cut in AWJ machining for 6063-T6 al alloy. In *Proceedings of world academy of science, engineering and technology,* vol. 59, pp. 142–145.

Kolahan F., Golmezerji R., Azadi Moghaddam M. (2012) Multi objective optimization of turning process using grey relational analysis and simulated annealing algorithm. *Applied Mechanics and Materials* 110:2926–2932. doi: 10.4028/www.scientific.net/AMM.110-116.2926.

Liao H. T., Chen Z. W. (2013) A study on fiber laser micro-spot welding of thin stainless steel using response surface methodology and simulated annealing approach. *International Journal of Advanced Manufacturing Technology* 67:1015–1025.

Manoochehri M., Kolahan F. (2014) Integration of artificial neural network and simulated annealing algorithm to optimize deep drawing process. *International Journal of Advanced Manufacturing Technology* 73(1-4): 241-249.

Mellal M. A., Williams E. J. (2016) Parameter optimization of advanced machining processes using cuckoo optimization algorithm and hoopoe heuristic. *Journal of Intelligent Manufacturing* 27:927–942. doi: 10.1007/s10845-014-0925-4.

Mukherjee I., Ray P. K. (2006) A review of optimization techniques in metal cutting processes. *Computers and Industrial Engineering* 50(1): 15–34.

Mukherjee I., Ray P. K. (2008) Optimal process design of two-stage multiple responses grinding processes using desirability functions and metaheuristic technique. *Applied Soft Computing* 8(1): 402–421.

Padmanaban R., Balusamy V., Saikrishna V., Gopath Niranthar K. (2014) Simulated Annealing Based Parameter Optimization for Friction Stir Welding of Dissimilar Aluminum Alloys. *Procedia Engineering* 97: 864 – 870. doi: 10.1016/j.proeng.2014.12.361.

Pawar P. J., Rao V. (2013) Parameter optimization of machining processes using teaching–learning-based optimization algorithm. *International Journal of Advanced Manufacturing Technology* 67:995–1006. doi: 10.1007/s00170-012-4524-2.

Rao R. V., & Pawar P. J. (2010). Parameter optimization of a multi-pass milling process using non-traditional optimization algorithms. *Applied Soft Computing,* 10(2), 445–456. ISSN 1568-4946, doi: 10.1016/j.asoc. 2009.08.007.

Rao R.V., Kalyankar V. D. (2014) Optimization of modern machining processes using advanced optimization techniques: a review. *International Journal of Advanced Manufacturing Technology* 73: 1159–1188.

Saravanan R., Asokan P., Vijayakumar K. (2003) Machining Parameters Optimisation for Turning Cylindrical Stock into a Continuous Finished Profile Using Genetic Algorithm (GA) and Simulated Annealing (SA). *International Journal of Advanced Manufacturing Technology* 21: 1–9.

Sathiya P., Aravindan S., Noorul Haq A., Paneerselvam K. (2009) Optimization of friction welding parameters using evolutionary computational techniques. *Journal of Materials Processing Technology* 209(5): 2576–2584.

Sibalija, T., Majstorovic, V., Miljkovic, Z. (2011) An intelligent approach to robust multiresponse process design, *International Journal of Production Research* 49(17): 5079-5097. doi: 10.1080/00207543.2010. 511476.

Sibalija T., Majstorovic, V. (2012a) An integrated approach to optimize parameter design of multi-response processes based on Taguchi method and artificial intelligence, *Journal of Intelligent Manufacturing* 23(5): 1511-1528. doi: 10.1007/s10845-010-0451-y.

Sibalija T., Majstorovic V. (2012b) An integrated, simulated annealing-based method for robust multiresponse process optimisation, International Journal of Advanced Manufacturing *Technology 59: 1227–12*44. doi: 10.1007/s00170-011-3572-3.

Sibalija T., Petronic S., Majstorovic V., Milosavljevic A. (2014). Modelling and optimisation of laser shock peening using an integrated simulated annealing-based method. *International Journal of Advanced Manufacturing Technology,* 73(5): 1141–1158. doi: 10.1007/s00170-014-5917-1.

Sibalija T., Majstorovic V. (2016) *The Advanced Multiresponse Process Optimization: an Intelligent and Integrated Approach*, Hardcover ISBN 978-3-319-19254-3, eBook ISBN 978-3-319-19255-0, Springer International Publishing Switzerland.

Somashekhar K.P., Mathew J., Ramachandran N. (2012) A feasibility approach by simulated annealing on optimization of micro-wire electric discharge machining parameters. *International Journal of Advanced Manufacturing Technology* 61: 1209–1213.

Spall J. (2003) *Introduction to Stochastic Search and Optimization*. John Wiley & Sons, Inc. New Jersey.

Sudhakaran R., Vel Murugan V., Sivasakthivel P. S., Balaji M. (2013) Prediction and optimization of depth of penetration for stainless steel gas tungsten arc welded plates using artificial neural networks and simulated annealing algorithm. *Neural Computing and Applications* 22:637–649. doi: 10.1007/s00521-011-0720-5.

Teimouri R., Baseri H. (2013) Artificial evolutionary approaches to produce smoother surface in magnetic abrasive finishing of hardened AISI 52100 steel. *Journal of Mechanical Science and Technology* 27 (2): 533-539. doi: 10.1007/s12206-012-1210-0.

Voß S. (2001). *Metaheuristics: The state of art*. Lecture Notes in Computer Science 2148: 1–23.

Yang S. H., Srinivas J., Mohan S., Lee D. M., Balaji S. (2009) Optimization of electric discharge machining using simulated annealing. *Journal of Materials Processing Technology* 209(9):4471–4475.

Yang S. H., Tzeng C. J., Yang Y. K., Hsieh M. H. (2012) Optimization of wire electrical discharge machining process parameters for cutting tungsten. *International Journal of Advanced Manufacturing Technology* 60:135–147. doi: 10.1007/s00170-011-3576-z.

Yen C., Lin J. C., Li W., Huang M. F. (2006) An abductive neural network approach to the design of runner dimensions for the minimization of warppage in injection mouldings. *Journal of Materials Processing Technology* 178: 1–8.

Yusup N., Zain A. M., Hashim S. Z. M. (2012) Evolutionary techniques in optimizing machining parameters: Review and recent applications (2007–2011). *Expert Systems with Applications* 39(10): 9909-9927.

Zain A. M., Haron H., Sharif S. (2010). Simulated Annealing To Estimate The Optimal Cutting Conditions For Minimizing Surface Roughness In End Milling Ti-6Al-4V. *Machining Science and Technology* 14: 43–62.

Zain A. M., Haron H., Sharif S. (2011a). Integration of simulated annealing and genetic algorithm to estimate optimal solutions for minimizing surface roughness in end milling Ti–6AL–4V. *International Journal of Computer Integrated Manufacturing,* 24(6): 574–592.

Zain A. M., Haron H., Sharif S. (2011b) Genetic algorithm and simulated annealing to estimate optimal process parameters of the abrasive water jet machining. *Engineering with Computers* 27(3): 251–259.

Zain A. M., Haron H., Sharif S. (2011c). Estimation of the minimum machining performance in the abrasive waterjet machining using integrated ANN-SA. *Expert Systems with Applications,* 38: 8316–8326.

Zandieh M., Amiri M., Vahdani B., Soltani R. (2009) A robust parameter design for multi-response problems. *Journal of Computational and Applied Mathematics* 230(2):.463–476.

BIOGRAPHICAL SKETCH

Tatjana Sibalija, PhD

Affiliation: Metropolitan University, Belgrade

Education: PhD Mech. Eng., BSc, MSc, PhD: Faculty of Mechanical Engineering, University of Belgrade, Serbia

Business Address: Tadeusa Koscuska 63, 11000 Belgrade, Serbia

Research and Professional Experience:

- Belgrade Metropolitan University, Belgrade, Serbia 2013 –
  Full Professor (Full Professor in Information Systems; Full Professor in Operations Management)
- Precise Corporation Co., Ltd., Bangkok, Thailand, 2013 -
  Technical director, Project: Industry 4.0 concept, infrastructure development, and implementation in Precise ltd, 2016-
  Technical director, Project: ICT-enabled Intelligent Manufacturing, 2013-2016
- European Commission, Brussels, European Union (Special Service Agreement), 2009 –FP7 / Horizon 2020 expert
- European University, Belgrade, Serbia 2010-2013
  Associate Professor
- Gazpromneft - Petroleum Industry of Serbia, Belgrade, Serbia 2008-2010
  Expert coordinator
- Faculty of Mechanical Engineering, University of Belgrade, Serbia, 2005-2008
  Lecturer and researcher
- ST Microelectronics, Malta, 2002-2005
  Engineer, manager
- Termoelektro-Projekt, Belgrade, Serbia, 2000-2002
  Design engineer, leading engineer

Honors:

- *Novum Quaels* – Annual award for outstanding contribution to science and technological development, 2015, Scientific association for development and affirmation of new technologies
- Outstanding Reviewer for *The TQM Journal*, 2015 Emerald Literati Network Awards for Excellence
- L. Romano Best Inventor Award (ST Worldwide Manufacturing Excellence), 2005
- ST Microelectronics, Malta -Technological Award, 2003

Publications from the Last 3 Years:

Book

1. Sibalija T., Majstorovic V. (2016) *The Advanced Multiresponse Process Optimisation: an Intelligent and Integrated Approach*, Hardcover ISBN 978-3-319-19254-3, eBook ISBN 978-3-319-19255-0, Springer International Publishing Switzerland

Journal papers

1. Sibalija T., Zivkovic S., Fountas N., Majstorovic V., Macuzic J., Vaxevanidis N. (2016) Virtual optimisation of CAI process parameters for the sculptured surface inspection, *Procedia CIRP*, Vol. 57, pp. 574–579, Elsevier, doi: http://dx.doi.org/10.1016/j.procir.2016.11.099
2. Petronic S., Sibalija T., Burzic M., Polic S., Colic K., Milovanovic D. (2016) Picosecond Laser Shock Peening of Nimonic 263 at 1064 nm and 532 nm Wavelength. *Metals* 2016, *6*(3), 41, doi:10.3390/met6030041
3. Stojadinovic S., Majstorovic V., Durakbasa N., Sibalija T. (2016) Towards an intelligent approach for CMM inspection planning of prismatic parts. *Measurement*, Vol. 92, pp.326–339, Elsevier, http://dx.doi.org/10.1016/j.measurement.2016.06.037
4. Marinkovic V., Bekcic S., Pejovic G., Sibalija T., Majstorovic V., Tasic Lj. (2016) An approach to TQM evaluation in pharma business, *The TQM Journal*, Vol. 2,8 Iss. 5, pp.745-759, Emerald, https://doi.org/10.1108/TQM-10-2015-0134
5. Stojadinovic S., Majstorovic V., Durakbasa N., Sibalija T. (2016) Ants colony optimisation of the measuring path of prismatic parts on CMM. *Metrology and measurement systems*, Vol. 23, Iss. 1, De Gruyter, doi: https://doi.org/10.1515/mms-2016-0011.

6. Majstorovic V., Markovic M., Kuzamovic D., Macuzic J., Marinkovic V., Sibalija T. (2015) ISO 9001:2015 — An Example of Application in Serbia. *Quality Assurance, Special Issue ISO* 9001:2015, ISSN: 1224-5410, Vol. XX1, No.83 (September 2015), pp.32-36.
7. Majstorovic V., Macuzic J., Sibalija T., Zivkovic S. (2015) Cyber-physical manufacturing systems – manufacturing metrology aspects, *Journal Proceedings in Manufacturing Systems*, Vol. 10, Iss.1, pp. 9-14, Romanian Academy Publishing House
8. Majstorovic V., Stojadinovic S, Sibalija T. (2015) Development of a knowledge base for the planning of prismatic parts inspection on CMM, *Acta Imeko*, Vol. 4, No.2, pp.10-17
9. Majstorovic V., Markovic M., Kuzmanovic D., Macuzic J., Marinkovic V., Sibalija T., Bekcic S., Pejovic G. (2015) Business standardization in Serbia and World - comparative study. Review paper, *Journal "Tehnika"*, Vol. 5/2015, pp. 884-892, doi: 10.5937/tehnika1505884M
10. Majstorovic V., Macuzic J., Sibalija T., Stojadinovic S., Zivkovic S. (2015) Horizon 2020 Programme for Industry 4.0 - Towards a digital model of quality, Review paper, *Journal "Tehnika"*, Vol. 2/2015, pp.376-382
11. Majstorovic V., Sibalija T. (2015) From IMS and six sigma toward TQM: an empirical study from Serbia, *The TQM Journal*, Vol. 27 Issue: 3, pp.341-355, Emerald, https://doi.org/10.1108/TQM-12-2013-0130
12. Majstorovic V., Sibalija T., Ercevic M., Ercevic B. (2014) CAI model for prismatic parts in digital manufacturing, *Procedia CIRP*, Vol. 25, pp.27-32, Elsevier, DOI: 10.1016/ j.procir.2014.10.006

Conference papers

1. Mijovic B., Sibalija T. (2017) Towards IoT-supported Business Process Management: a Case Study, *Proceedings of the First International Student Scientific Conference "Multidisciplinary Approach to Contemporary Research,"* Belgrade, 25.-26. November 2017
2. Jovanovic S., Sibalija T. (2017) The role of business process management and its implementation in telecommunication sector using SAP methodology, *Proceedings of the First International Student Scientific Conference "Multidisciplinary Approach to Contemporary Research,"* Belgrade, 25.-26. November 2017
3. Sibalija T., Punovic V. (2017) On the use of machine learning techniques in e-learning systems, Proceedings of The 8th International *Conference on eLearning (eLearning-2017)*, Belgrade, Serbia, 28-29 September 2017
4. Ljubic A., Sibalija T. (2017) Analysis of soft factors in Six Sigma implementation, *Proceedings of the International Conference "Innovation, Competitiveness and Sustainable Development," ICSD* Belgrade 2017**,** Belgrade, Serbia, 25. May 2017
5. Sibalija T. (2016) COBIT 5: A comprehensive framework for IT processes assessment and governance in a digital enterprise, *Proceedings of the 3rd International Conference "Modern Methods of Testing and Evaluation in Science,"* Belgrade, Serbia, December 2016
6. Sibalija T. (2016*)* Digital factory: the information systems aspect, Invited Keynote paper/Introduction paper, *Proceedings of the 13th International Convention on Quality - UASQ 2016*, ISBN 978-86-89157-07-9, pp. 1-6, United Association of Serbia for Quality (UASQ), Belgrade, 31.May – 3.Jun 2016
7. Ljubic A., Sibalija T. (2016) Business process management and Six Sigma: a synergy for process-oriented organisation, *Proceedings of the 13th International Convention on Quality -*

UASQ 2016, ISBN 978-86-89157-07-9, pp. 7-12, United Association of Serbia for Quality (UASQ), Belgrade, 31. May – 3.Jun 2016

8. Dimitrijevic M., Ljubic A., Sibalija T. (2016) Implementation of business process reengineering and DMAIC in manufacturing sector: a case study, *Proceedings of the 13th International Convention on Quality - UASQ 2016*, ISBN 978-86-89157-07-9, pp. 13-18, United Association of Serbia for Quality (UASQ), Belgrade, 31. May – 3. Jun 2016
9. Sibalija T. (2015) Cyber-security in digital manufacturing: assessment and testing, *Proceedings of the 2nd International Conference "Modern Methods of Testing And Evaluation in Science,"* Belgrade, Serbia, 14. December 2015.
10. Majstorovic V., Macuzic J., Marinkovic V., Sibalija T. (2015) Application of ISO 9001:2015. *World Quality Forum "Quality for Future of the World," International Academy for Quality and Hungarian National Committee for EOQ*, Budapest, Hungary, 26.-27. October 2015.
11. Majstorovic V., Macuzic J., Stojadinovic S., Zivkovic S., Sibalija T., Marinkovic V. (2015) *Cyber Physical Manufacturing – Integrated Quality Approach. Proceedings of 6th International Symposium of Industrial Engineering – SIE* 2015, pp. 137-140, ISBN 978-86-7083-864-2, University of Belgrade, Faculty of Mechanical Engineering, Belgrade, Serbia, 24.-25. September 2015.
12. Fountas N., Sibalija T., Majstorovic V., Vaxevanidis N., Macuzic J., Zivkovic S. (2015) Virtual quality assessment for sculptured surface CNC tool path strategies and related parameters using RSM and generated DMIS file, Keynote paper, *Proceedings of the 8th International Conference "Total Quality Management-Advanced and Intelligent Approaches,"* Belgrade, 1.-5. June 2015.

13. Marinkovic V., Sibalija T., Bekcic S, Pejovic G, Majstorovic V., Tasic Lj. (2015) TQM in pharmaceutical industry in Serbia, Keynote paper, *Proceedings of the 8th International Conference "Total Quality Management-Advanced and Intelligent Approaches,"* Belgrade, 1.-5. June 2015.
14. Sibalija T. (2014) Novel method for multiresponse optimisation based on simulated annealing algorithm, *Invited plenary* paper, *Proceedings of the 1st National Conference "Modern Methods of Testing and Evaluation in Science,"* ISBN 978-86-918415-0-8, pp.2-12, Belgrade, 25.December 2014.

In: Simulated Annealing
Editors: Alex Scollen et al.
ISBN: 978-1-53613-674-6

Chapter 2

USE OF SIMULATED ANNEALING FOR BIOPHYSICAL ANALYSES OF BIOLOGICAL MACROMOLECULAR SYSTEMS

Jiyoung Kang[1], Takuya Sumi[2], Hiroshi Yamaguchi[3] and Masaru Tateno[2,*]

[1]Institute of Human Complexity and Systems Science,
System Science Center for Brain and Cognition,
Yonsei University, Seoul, Republic of Korea
[2]Graduate School of Life Science,
University of Hyogo, Hyogo, Japan
[3]Nagoya University Brain & Mind Research Center,
Aichi, Japan

ABSTRACT

We describe applications of the simulated annealing (SA) method to biological macromolecular systems. The first example is the reconstruction of an inappropriate three-dimensional (3D) structure of a functional site in

---

* Corresponding author: Email: tateno1611@gmail.com.

a protein, built based on X-ray crystallographic experiments. The SA method significantly extended the sampling space of the conformational space, and thus enabled the identification of the energetically-optimal 3D-structure of the functional site in the enzyme structure. The obtained structure facilitated various biophysical analyses, thereby leading us to theoretically resolve a discrepancy found in the biochemical experimental data.

We also briefly introduce another example of the use of the SA method to resolve structural issues, by constituting an extremely advanced energy function (Hamiltonian) for computer simulations of biological macromolecular systems. The SA method enhanced the explorations of the optimal parameter set defined in such an energy function, so its values could be fit with those obtained by the exclusively high level *ab initio* quantum mechanics calculations. This type of energy function (*i.e.*, an effective function) is very important to precisely describe molecular interactions (e.g., stacking interactions) that are frequently found and critical in the drug design field.

In an application to bioinformatics field, the SA scheme was combined with the Multivariate Curve Resolution-Alternating Least Squares (MCR-ALS) method, for the first time (our scheme is referred to as the SA-MCR-ALS method). The conventional MCR-ALS has been widely employed to analyze various multivariate data, such as spectrum data from circular dichroism (CD) and nuclear magnetic resonance (NMR) analyses, and image data from bioinformatics fields. However, it still fails to find the appropriate solutions in some cases. We describe an example where the MCR-ALS method fails, and how the SA-MCR-ALS method overcomes those difficulties.

In this manner, the SA scheme is a fundamental and powerful method that is intensively applicable to resolve the local minimum problem in various scientific fields related to biological macromolecular systems. In this chapter, we demonstrate such examples together with introductions of the relevant various technologies employed with the SA method.

Keywords: computational biology, theoretical biology, structural biology, rational drug design, molecular dynamics (MD) simulation, density functional theory (DFT), CCSD(T) at a basis set limit level, multivariate analysis, DNA, peptide, protein

1. Introduction

The simulated annelining (SA) method is a fundamental scheme used to resolve the local minimum problem in various scientific fields. In this chapter, some applications of the SA method to biological macromolecular systems are introduced. The first application is the reconstruction of an inappropriate three-dimensional (3D) structure of a functionally-important site in a protein enzyme structure obtained by X-ray crystallographic data (Kang, Kino, & Tateno, 2011). The SA method significantly extended the sampling space of the conformational space, and thus allowed the identification of the energetically-optimal 3D-structure of the target site in the protein structure (Kang, Kino, & Tateno, 2011).

Employing the obtained structure, we conducted *ab initio* electronic structure calculations of the functional site within a framework of hybrid *ab initio* quantum mechanics (QM)/molecular mechanics (MM) calculations (Kang, Kino, & Tateno, 2011), which enabled us to perform *ab initio* electronic structure calculations of the target site (corresponding to the QM region), including the "entirety" of the "huge" enzyme structure (the remaining part is assigned to the MM region), without increasing the computational costs as compared to the "isolated" structure corresponding to the QM region (Hagiwara, Ohta, & Tateno, 2009; Hagiwara & Tateno, 2010; Kang et al., 2009). Employing these simulations, we analyzed the geometrical and electronic structures of the target site, and thereby resolved a discrepancy that was found in previous biochemical experimental data (Kang, Kino, & Tateno, 2011).

For structural issues of biological macromolecular systems, we also introduce crucial applications of the SA method to the computational rational drug design fields. It is very important to constitute an extremely advanced energy function (Hamiltonian) for high-quality computer simulations of biological macromolecular systems. The SA method enhanced the explorations of the optimal parameter set defined in such an energy function so as to reproduce extremely superior-level *ab initio* electronic structure calculations. This type of highly advanced energy function is essential to precisely describe particular molecular interactions, such as stacking interactions, that are frequently found and critical in the drug design field.

In the second part, the SA method was combined with the Multivariate Curve Resolution-Alternating Least Squares (MCR-ALS) scheme, and thereby we developed a novel method, the SA-MCR-ALS scheme (Kang et al., 2017). The conventional MCR-ALS has been widely employed to analyze various multivariate data, such as the spectrum data from circular dichroism (CD) and nuclear magnetic resonance (NMR) analyses, and the image data from bioinformatics fields. However, MCR-ALS still fails to find the appropriate solutions in some cases. To solve this problem, we first show an example where the conventional MCR-ALS failed. When it was used to process the CD spectra of double-stranded DNA (228 bp) in the complex with various concentrations of a DNA-binding peptide, only unreasonable solutions were provided. To overcome this failure, we developed the SA-MCR-ALS method, and successfully decomposed the spectral data into the pure component spectra. This achievement enabled us to investigate the interactions between the DNA and the peptide, thus leading to structural modeling of the DNA-peptide complex at atomic resolution.

In this manner, the SA-based methodologies significantly support scientific work in various biological fields. Thus, in this chapter, we review such studies, together with introductions of the relevant various biophysical techniques employed with the SA method.

2. Reconstruction of the 3D Structure of a Protein

X-ray crystallographic structural analyses are widely used to determine the 3D structures of biological macromolecular systems. However, in the transition metal-binding site, the electrons are delocalized, and the intensive X-ray beam reduces the functional site. Thus, the 3D structural modeling employing X-ray crystallographic data occasionally fails to assign the appropriate and accurate structures. Moreover, the crystal packing may also induce distortions of the 3D-structures that are different from those of the physiological conditions. For these reasons, explorations of the optimal geometric structures of such biological macromolecular systems are essential to understand their functional mechanisms. In this section, we first show an inappropriate 3D structure identified in a crystallographic modeled structure, and then describe how we explored the optimal structure using the SA method.

We discovered that the crystal structure of an enzyme contained a structural discrepancy in a functional site that was known to be relevant to the electron transfer from another protein, although the resolution of the crystal structure was relatively higher (1.9 Å) as a biological system. The enzyme is cytochrome c oxidase (CcO), and the functional site that we analyzed as our target in the enzyme is the Cu_A site, which contains two Cu ions and six Cu-coordinated amino acid residues (Figure 1). The two Cu ions are commonly coordinated by the S atoms of two cysteine residues (Cys196 and Cys200). A Cu ion, Cu_O, is coordinated by an N atom (of the His204 side chain) and a carbonyl O atom (of the Glu198 backbone), and the other Cu ion, Cu_S, is coordinated by an N atom (of the His161 side chain) and an S atom (of the Met207 side chain).

Notably, as inferred from the Cu_O-Cu_S atomic distance, these two Cu ions form a covalent bond, and thereby the electrons of the two Cu ions are highly shared. Thus, the structure of the Cu_A site is stabilized by the covalent bond between the Cu ions and the coordinated amino acid residues. However, such delocalization of the electron densities may increase the difficulty of the crystallographic modeling; i.e., the construction of the appropriate 3D structure based on the electron densities obtained by the X-ray crystallographic experiments. Actually, the atomic distance between two carbon atoms, C_β (Cys196)-C_γ (Met207), in the Cu_A site, was too short (3.40 Å), and the shortest distance between two hydrogen atoms was as small as 1.67 Å, when hydrogen atoms were attached to the crystal structure. Thus, we found that the structure provided by the crystallographc modeling was energetically unstable (Figure 1).

In order to efficiently search for the optimal 3D-structure, we established a calculation scheme, as follows. In the first step, we attached the hydrogen atoms to the crystal structure of bovine cytochrome c oxidase (PDB ID: 2EIJ) (Muramoto et al., 2007), and the positions of all of the attached hydrogen atoms were optimized by energy minimizations and multiple-step MD simulations coupled with our SA scheme. Specifically, we conducted the MD simulations of the Cu_A site in cytochrome c oxidase at various temperatures, at which only the hydrogen atoms were mobile. In this stage, the inappropriate conformations of the Cys196 and Met207 residues were not relaxed yet, and thus further optimizations of the Cu_A structure were required.

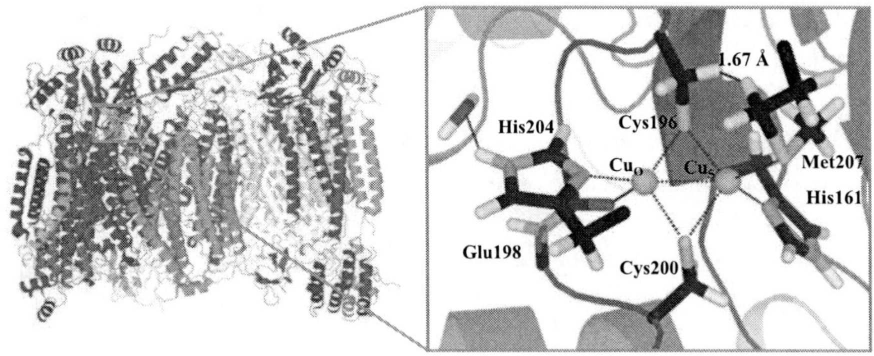

Figure 1. Crystal structure of the Cu_A site of bovine cytochrome c oxidase. The initial coordinates of the protein were obtained from the fully-reduced crystal structure of bovine CcO (PDB ID: 2EIJ) (Muramoto et al., 2007). Reproduced from *Biochim. Biophys. Acta*, *1807* (10), Kang, J., Kino, H. & Tateno, M., A theoretical investigation of the functional role of the axial methionine ligand of the Cu_A site in cytochrome c oxidase, 1314-1327, Copyright (2011), with permission from Elsevier.

In the second step, we explored the conformations of Cys196 and Met207 (in which all of the atoms were mobile) with energy minimizations and multiple-step MD simulations with our SA scheme (i.e., the MD calculations were performed at various temperatures) (Figure 2). In the sequential multiple-step MD simulations, the calculations were started from a lower temperature (200 K), and then the temperature was gradually increased (toward 500 K). Subsequently, the system was cooled down toward the lower temperature (200 K). In the MD simulations at higher temperatures, the system can easily overcome the energy barriers and escape from the local minima. However, in the MD simulations at lower temperatures, a careful conformational search can be performed. In this manner, we obtained a relatively stable conformation by our SA extended sampling method.

In the third step, we set a water sphere with a 30 Å radius, to consider the solvent effect (1,064 additional water molecules were arranged to solvate the Cu_A site). The structure of the ensemble composed of the solvent water molecules was carefully relaxed by our protocol. (The three step procedure, involving a 50-step energy minimization, a 10-ps MD simulation, and a 50-step energy minimization, was conducted to optimize the positions of the hydrogen atoms of the water molecules, and then the positions of the water molecules themselves. Finally, a 30-step energy minimization was performed for the entire system).

In the final (fourth) step, employing the obtained structure, we performed hybrid *ab initio* quantum mechanics (QM)/molecular mechanics (MM) calculations, by which we explicitly considered the electronic structures of the Cu_A site. Due to the high computational cost of the *ab initio* QM calculations, it is too difficult to analyze the whole structure of the biological macromolecular system as the QM region. However, hybrid *ab initio* QM/MM calculations can actualize the *ab initio* electronic structure calculations of the functional and active regions (which should be assigned to the QM region) within a huge macromolecular system (the rest of the atoms should be assigned to the MM region), and thus it is a state-of-the-art method to investigate the electronic and geometric structures of those systems. More specifically, in the hybrid *ab initio* QM/MM method, the functional site, where the electronic structural changes are important, is considered as the QM region, and the other part of the enzyme, the solvent water molecules, and the membrane parts are assigned to the MM region. We constructed our computational system to conduct hybrid *ab initio* QM/MM calculations on supercomputers (Hagiwara, Ohta, & Tateno, 2009).

In this manner, the electronic and geometric structures of the functional site were obtained with the consideration of the environmental effects. Even though the QM/MM method allows us to conduct efficient calculations, the cost of the QM part is still huge, and so the accurate initial structure for the hybrid *ab initio* QM/MM simulation is important. Thus, the SA extended sampling framework performed in the second stage is very effective, especially when coupled to other simulations, such as hybrid *ab initio* QM/MM calculations.

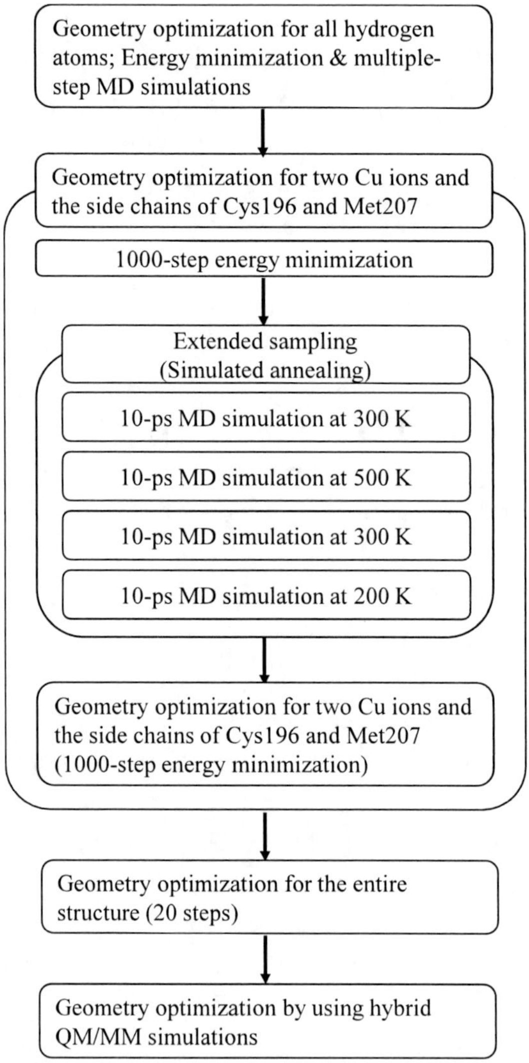

Figure 2. Calculation scheme of our SA extended sampling method to search the conformational space.

In our hybrid *ab initio* QM/MM simulations of CcO (note that the entire enzyme structure was included in the simulations), two Cu ions, six coordinated amino acid residues (all coordinated to the Cu ions), and one crystal water molecule were assigned to the QM region. The other parts of the protein and the solvent water molecules were assigned to the MM region. For the hybrid *ab initio* QM calculation, the density functional theory (DFT) calculation method was adopted, where the B3LYP functional, which is astonishingly adaptable to a broad range of materials, including biological macromoleucles, was employed for the DFT calculation (Becke, 1993).

Thanks to the combination of the efficient SA extended conformational sampling and the sophisticated hybrid *ab initio* QM/MM simulations, we successfully resolved a discrepancy that was found in the experimental data. In fact, the obtained structure was energetically stable, and fit very well to the electron density map (Figure 3A). Furthermore, the singly occupied molecular orbital (SOMO) obtained by our hybrid *ab initio* QM/MM simulation provided an explanation for the characteristic electronic structure (the mixed valence state) of the Cu_A site (Figure 3B).

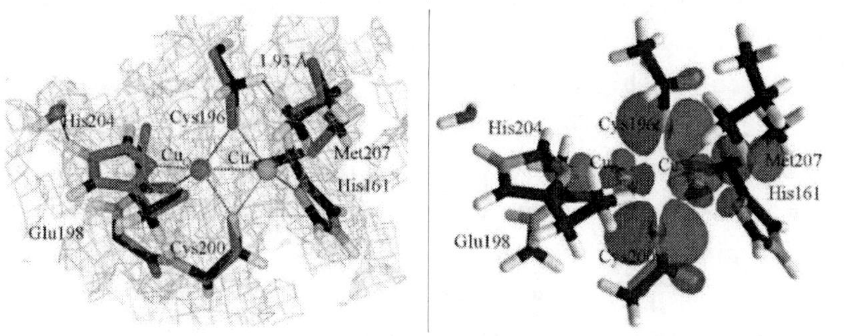

(A) (B)

Figure 3. The optimized 3D structure (A) and the SOMO (B) of the Cu_A site, obtained by the combination of efficient SA extended conformational sampling and sophisticated hybrid *ab initio* QM/MM simulations. The electron density map is shown by the green mesh. In the optimized structure, the atomic distance between Hβ of Cys196 and Hγ of Met207 was 1.93 Å, indicating the reduction of the steric clash identified in the crystal structure. Reproduced from *Biochim. Biophys. Acta*, *1807* (10), Kang, J., Kino, H. & Tateno, M., A theoretical investigation of the functional role of the axial methionine ligand of the Cu_A site in cytochrome c oxidase, 1314-1327, Copyright (2011), with permission from Elsevier.

Based on the obtained structure, we investigated the functional roles of the amino acid residues in the Cu_A site, by employing hybrid *ab initio* QM/MM simulations. In the subsequent part, we briefly introduce our biophysical and structural analyses using the refined structure obtained by our SA extended conformational sampling method.

CcO is the terminal enzyme of the respiration chain, and it reduces an oxygen molecule to two water molecules, thereby generating a gradient of protons between the mitochondrial matrix and intermembrane space. Another enzyme, cytochrome c (Cyt c), transports the electrons required for this biochemical process. All of the transported electrons enter into the Cu_A site of CcO, and these electrons are sequentially transferred to heme *a*, heme a_3, and the Cu_B site (Farver, Grell, Ludwig, Michel, & Pecht, 2006; Moser, Page, & Dutton, 2006; Papa, Capitanio, & Capitanio, 2004; Regan, Ramirez, Winkler, Gray, & Malmstrom, 1998). Accordingly, the Cu_A site is considered to be the "gate" of the electrons required for the reaction of CcO.

Although it was related to the redox potential, the functional role of Met207, which is coordinated to the Cu_S ion, was unknown. Actually, in biochemical studies attempting to elucidate the role of Met207, an experimental discrepancy was found, as follows. A mutant of the engineered azurin, which was structurally altered so it possessed the Cu_A site, suggested that the defect of Met207 only slightly affected the electron transfer activity (Hwang, Berry, Nilges, & Lu, 2005; Robinson et al., 1999). In contrast, a similar mutant of *Rhodobacter sphaeroides* CcO showed that the mutation of Met207 significantly changed the redox potential of the Cu_A site (Wang et al., 2002; Zhen, Schmidt, Kang, Antholine, & Ferguson-Miller, 2002). Therefore, the conclusions were completely different, although the 3D-structures of the Cu_A sites in these enzymes were quite similar to each other.

In our study employing hybrid *ab initio* QM/MM simulations of CcO involving the Cu_A site, for which the structure was obtained by our SA extended conformational sampling method, we solved the aforementioned experimental discrepancy; i.e., the apparently different effects of Met207 depending on the species of the enzymes, as follows (Kang, Kino, & Tateno, 2011). To analyze the effect of Met207, we constructed various mutants of the Cu_A site by employing computer modeling techniques, and then performed hybrid *ab initio* QM/MM simulations of the mutants. In the comparison of the electronic structures between the wild type and mutants of Met207, we found that Met207 did not induce significant changes of the electronic structure of the Cu_A site, except for the following one: In a mutant of Met207, the energy of a particular molecular orbital (MO) that might be related to the electron transfer increased, while the changes in the energy levels of the other MOs were marginal.

Furthermore, we performed homology modeling of the Cu_A site, and constructed the enzyme structures of the other aforementioned species. The analyses revealed that for the *Rhodobacter sphaeroides* enzyme, a large redox potential change was induced, which may originate from the insertion of the Ser155-Thr196 region. Thus, we identified an inserted structural element close to the Cu_A site as a functional segment, and furthermore suggested that Met207 serves as a fine modulator that can assist in adaptations to various environments, depending on the species.

Note here again that the Cu_A site involves the two Cu ions (Cu_O and Cu_S), two S^- ions (S), and two imidazole rings (Im), and thus the fundamental structure is represented by $Cu_2S_2Im_2$, which can be constructed in a plane. This moiety forms the fundamental framework of the electronic structure of the Cu_A site, which is responsible for the electron transfer. In this context, Met207 contributes to the fine regulation of the functions of the Cu_A site.

In summary, our SA scheme, combined with MD simulations, significantly extended the sampling space of the conformational space, and thus allowed us to identify the energetically optimal 3D-structure of the Cu_A site in CcO. Employing the obtained structure, we conducted *ab initio* electronic structure calculations of the Cu_A site within a framework of hybrid *ab initio* QM/MM calculations, and thus resolved the discrepancy found in the biochemical experimental data. Based on further hybrid *ab initio*

QM/MM simulations coupled to the structural modeling of the mutants of Met207, we revealed that the inserted structural elements provided an explanation for the functional effects of the Met207 residue that were different in the *R. sphaeroides* enzyme and the engineered azurin.

In this section, we showed how the modeled structure based on the crystallographic experimental data can be improved and refined by the SA conformational sampling method coupled to MD simulations and hybrid *ab initio* QM/MM calculations. The advantages of such extended structural sampling can also be applied to drug design (Anderson, 2003; Macalino, Gosu, Hong, & Choi, 2015; Sliwoski, Kothiwale, Meiler, & Lowe, 2014). For example, AutoDock, a software suit for docking drugs to receptors, adopted the SA method to generate various conformations of ligands (Goodsell, Morris, & Olson, 1996). Another program for drug docking to receptors, the multi-conformation simulated annealing-pseudo-crystallographic refinement (MCSA-PCR), also utilizes the SA method to overcome the local minimum problem (Ota & Agard, 2001). Moreover, Cheng et al. combined the SA method and the genetic algorithm for the drug design (Chien-Cheng, Leuo-Hong, Cheng-Yan, Ming, & Wen-Chin, 1998). In this manner, the SA method has already been proven to be an important and powerful scheme to efficiently search the conformational and configurational spaces of drugs in docking examinations with receptors.

Use of the SA Method toward Computational Rational Drug Design

As another application of the SA method toward drug design, we also briefly provide an example from our previous studies, which is the construction of a sophisticated and advanced energy function (Hamiltonian) as an effective energy function employed for classical MD simulations. We need such an advanced energy function in the framework of classical mechanics when the biological macromolecular systems include particular interactions, such as stacking interactions. These molecular interactions are frequently found in interactions between drugs and receptors (proteins) (Figure 4A). Even if we employed *ab initio* electronic structure calculations for these systems, it would be very difficult to precisely describe such interactions without applying highly advanced *ab initio* calculation methodologies, as described below. In addition, the calculation costs of the highly advanced *ab initio* electronic structure calculations are too expensive to perform long-term MD simulations that are directly coupled with this highly advanced methodology. Accordingly, to overcome these difficulties, the development of an effective (potential) energy function is very important for the efficiency, as described below.

Stacking interactions, such as π-π, π-cation, and π-water interactions, have been widely observed in various biological macromolecular systems (Figure 4), and are known to contribute to the thermodynamic stabilization (Cerny & Hobza, 2007; Chan, Prenner, & Vogel, 2006; Dougherty, 1996, 2007; Gallivan & Dougherty, 1999; Ma & Dougherty, 1997; Meyer, Castellano, & Diederich, 2003; Minoux & Chipot, 1999). Although the current standard force field parameters for MD simulations have been intensively improved, the stacking interactions have not been evaluated correctly. Considerations of the van der Waals (vdW) energy are important to evaluate stacking interactions with a high accuracy (Cerny & Hobza, 2007; Tsuzuki, Honda, Uchimaru, Mikami, & Tanabe, 2002). However, including the vdW energy in *ab initio* electronic structure calculations still requires sophisticated and advanced *ab initio* schemes, such as Quantum Monte Carlo (QMC) calculations and coupled cluster with singles, doubles, and perturbative triples (CCSD(T)) methods (Anderson, 1976; Head-Gordon, Pople, & Frisch, 1988; Pople, Head-Gordon, & Raghavachari, 1987). The computational costs of these advanced methods are huge, and thus it is currently impossible to employ them in MD simulations of biological macromolecular systems.

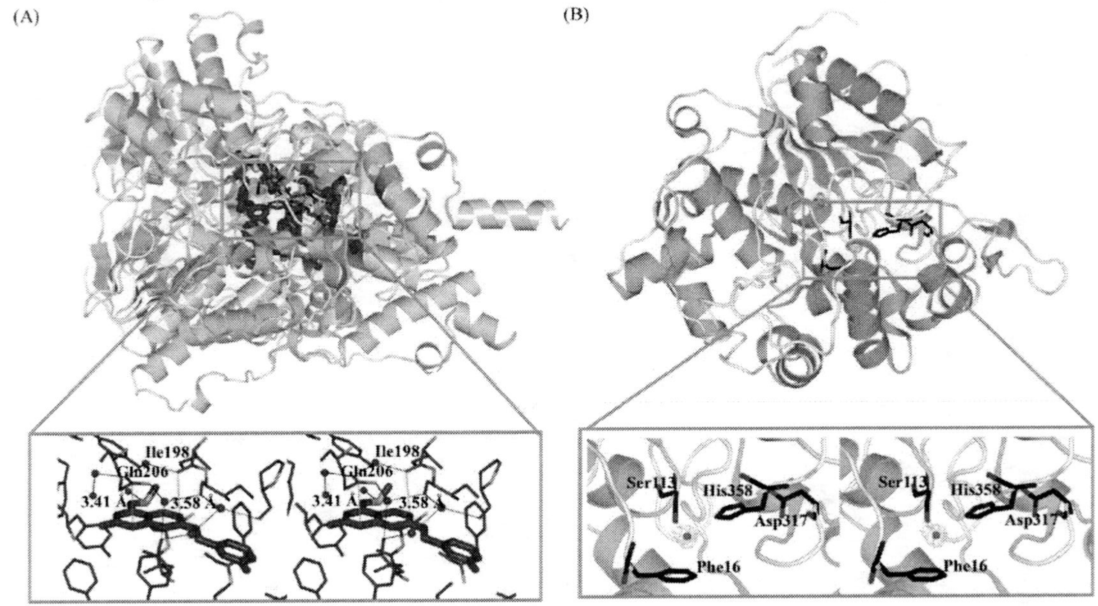

Figure 4. (A) The crystal structure of monoamine oxidase B (MAOB) (PDB ID: 2V61) (Binda et al., 2007). As shown in the lower panel (stereo view), σ-π interactions represented by green broken lines were identified between the ligand (7-[(3-chlorobenzyl)oxy]-4-[(methylamino)methyl]-2h-chromen-2-one) and amino acid residues (Ile198 and Gln206) in the catalytic site of MAOB. Note that the ligand in the complex is a drug medically employed for imaging by positron emission tomography (PET) in clinical examinations. (B) The crystal structure of T1 lipase (PDB ID: 2DSN) (Matsumura et al., 2008). As shown in the lower panel (stereo view), a cation-π interaction was identified as the Na^+-π structural element in the catalytic site of T1 lipase. Our theoretical analysis, employing the GER energy function, indicated that this Na^+-π element is important to maintain the structural stability.

In our efforts to overcome this problem, we developed the spatial grid-based energy representation (GER), which can reproduce the energy function (i.e., the force field parameters) with high accuracy at the basis-set limit CCSD(T) level, but with a reasonable computational cost almost identical to that of the classical MD simulation (Hagiwara & Tateno, 2009). To construct the energy function based on the GER framework, we employ our SA method and obtain an appropriate parameter set in the GER energy function. Actually, by employing the obtained parameter set, we constituted a GER energy function and applied it to MD simulations of T1 lipase, in order to evaluate the cation-π interaction within the enzyme (Hagiwara, Kang, & Tateno, 2011; Hagiwara, Matsumura, & Tateno, 2009) (Figure 4B).

The current standard energy function for a classical MD simulation consists of two parts, the bonded and non-bonded interactions. In the energy function of the bonded interactions, the covalent bonds are modeled as the harmonic oscillators. For the non-bonded interactions, the energy function includes the electrostatic and vdW interactions. Here, the total energy function of a standard energy function for the MD simulation is

$$E_{total} = E_{bonded} + E_{Coulomb} + E_{vdW} . \tag{1}$$

The vdW interaction in the Amber force field is given by the sum of the repulsive and attractive parts, as follows:

$$E_{\mathrm{vdW}}^{\mathrm{Amber}} = \sum_{i<j} \left(\frac{A_{ij}}{R_{ij}^{12}} - \frac{B_{ij}}{R_{ij}^{6}} \right). \tag{2}$$

In the GER, for the stacking interaction, we employed the Andersson-Langreth-Lundqvist (ALL) functional to improve the attractive vdW interaction term of the conventional force field (Andersson, Langreth, & Lundqvist, 1996). The improved attractive vdW interaction between stacked two molecules is given as follows:

$$E_{\mathrm{vdW}}^{\mathrm{ALL}} = \frac{-6}{(4\pi)^{3/2}} \int_{V_1}\int_{V_2} \frac{\rho^{1/2}(\vec{r}_1)\rho^{1/2}(\vec{r}_2)}{\rho^{1/2}(\vec{r}_1)+\rho^{1/2}(\vec{r}_2)} \frac{1}{r_{12}{}^{6}} d\vec{r}_1^{\,3} d\vec{r}_2^{\,3} \; . \tag{3}$$

The modified total energy is given as follows:

$$E = E_{bonded} + \sum_{i<j}\left(\frac{q_i q_j}{\varepsilon R_{ij}} + \frac{A_{ij}}{R_{ij}^{12}} \right) - E_{vdw}^{ALL} \; , \tag{4}$$

where the electron density $\rho(r)$ is represented as follows:

$$\rho(\vec{r}) = \sum_{i}^{N_{\mathrm{atoms}}} q_i a_{i,x} \exp\left(-a_{i,x}^2 (x-x_i)^2\right) a_{i,y} \exp\left(-a_{i,y}^2 (y-y_i)^2\right) a_{i,z} \exp\left(-a_{i,z}^2 (z-z_i)^2\right) . \tag{5}$$

The parameters in the GER function (i.e., $a_{i,x}$, $a_{i,y}$ $a_{i,z}$) must be fitted to reproduce an accurate energy function. We adopted an SA method to fit these parameters of the GER function (Figure 5A).

To obtain highly accurate energy profiles, we employed an extrapolation method proposed by Tsuzuki et al. (Tsuzuki, Uchimaru, & Mikami, 2006). According to this method, the energy of stacking interactions can be calculated, with the CCSD(T) at a basis set limited level, in which the basis set used for the calculations is considered as the basis set with infinite dimensions. Thus, the accuracy of the CCSD(T) calculation at a basis set limit level is quite high. However, the calculation cost of our GER energy function is comparable to that of the standard MD simulation, while the accuracy of the energy calculation is equivalent to that of the CCSD(T) at a basis set limited level.

Thus, our methodology to constitute the GER energy function coupled to our SA scheme confers great advantages, in terms of both accuracy and calculation cost. The obtained GER energy profile accurately reproduced that calculated by our CCSD(T), at the basis set limit level (Figure 5B). Actually, this type of highly advanced energy function is essential for drug design, as mentioned above, when the molecular system includes stacking interactions. In this manner, the SA method has been indispensable in the field of biological physics.

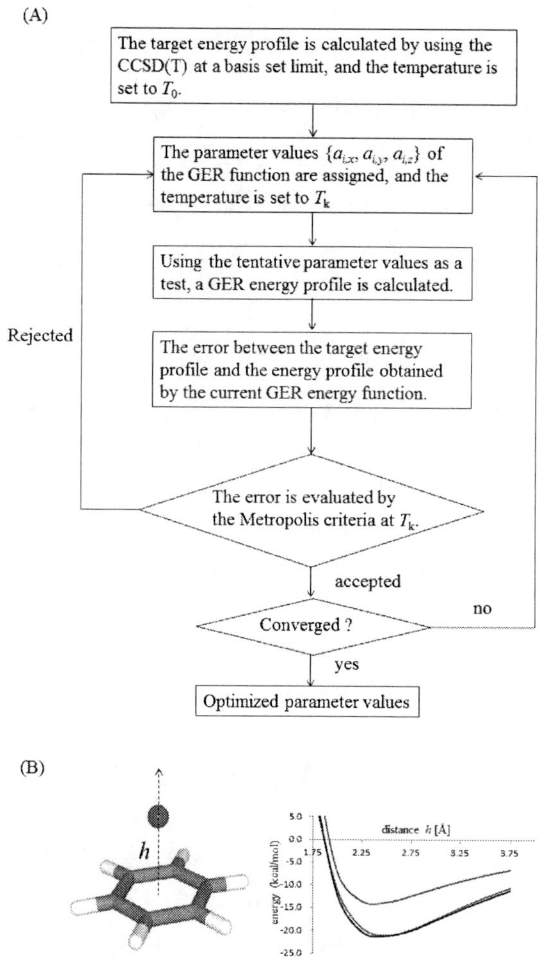

Figure 5. Calculation scheme of the SA extended sampling method to search in the parameter space of the GER energy function (A), and the generated GER energy function created for the Na^+-benzene interaction (B). The energy profiles of the GER, CCSD(T), and a standard force field for MD simulation (ff99) are represented by black, blue, and red lines, respectively. Reproduced (adapted) with permission from *J. Am. Chem. Soc., 131* (46), Hagiwara, Y., Matsumura, H. & Tateno, M., Functional Roles of a Structural Element Involving Na+-π Interactions in the Catalytic Site of T1 Lipase Revealed by Molecular Dynamics Simulations, 16697-16705. Copyright (2009) American Chemical Society.

3. SA-MCR-ALS

Recently, multivariate experimental data have increased in various scientific and engineering fields, including chemical and biological physics, highlighting the importance of the computational analysis of such multidimensional data (Felten et al., 2015; Garrido, Rius, & Larrechi, 2008; Haeusler et al., 2014; Jaumot, Gargallo, de Juan, & Tauler, 2005; Jaumot & Tauler, 2010; Maizels, 2015; Meshki, Behpour, & Masoum, 2015; Olmos et al., 2017; Ruckebusch & Blanchet, 2013; Wentzell et al., 2006; Wolffe, 2001). In this final section, we introduce an application of our SA scheme to extend the Multivariate Curve Resolution Alternating Least-Squares (MCR-ALS) method, which is currently popular for data analysis in spectroscopic and bio-imaging techniques. This method minimizes the fitting error under some constraints, such as non-negativity, unimodality, and closure, concerning either the spectral or concentration profiles, which are pre-knowledge of the physically meaningful information. Although the MCR-ALS calculation is now widely used to process multivariate curves, such as those obtained by infrared absorption, circular dichroism (CD), and NMR analyses, as well as various image data (Felten et al., 2015; Garrido, Rius, & Larrechi, 2008; Jaumot, Gargallo, de Juan, & Tauler, 2005; Jaumot & Tauler, 2010; Meshki, Behpour, & Masoum, 2015; Ruckebusch & Blanchet, 2013; Wentzell et al., 2006), it occasionally falls into a local minimum.

Here, we present such an example. We first conducted the conventional MCR-ALS analysis for our CD spectroscopic data that indicated the conformational transitions of double-stranded DNA (dsDNA), induced by the binding of various concentrations of a DNA-binding peptide (this peptide was previously reported to exhibit a preference for binding supercoiled DNA) (Tsutsui, Sano, Hosoya, Miyamoto, & Tsutsui, 2011). The conventional MCR-ALS analysis produced unreasonable spectral profiles. For example, some of the spectral components were complementary to each other, although the number of spectral components was likely to be reasonable. To resolve the local minima problem of MCR-ALS, we combined a simulated annealing (SA) protocol with the MCR-ALS method, which we refer to as the SA-MCR-ALS method. The SA-method allows quite effective exploration of the sampling space. By employing the SA-MCR-ALS scheme, the CD-spectral data of the dsDNA were successfully decomposed, and thus the spectral profiles and concentrations of distinct DNA conformations were identified for various conditions, including different peptide concentrations.

In this section, we first introduce the SA-MCR-ALS calculation scheme, and then describe its application to the CD spectra of the complex of the dsDNA and the peptide. Finally, we discuss the meaning of this analysis coupled with the SA method, from a biophysical point of view.

In the MCR-ALS theory, the data matrix D is decomposed as follows:

$$D = CS^T + E. \quad (6)$$

Matrices C and S^T represent the concentrations of the spectral factors and the pure spectra profile, respectively. The superscript T represents the transpose of the matrix, and the residual matrix E represents the experimental error.

The dimensions of the data matrix D are $N_r \times N_w$, where N_r and N_w are the concentrations of the peptide and the measured points of the wavelengths, respectively. The number of spectra N_s determines the dimensions of C and S^T; i.e., C and S^T are the $N_r \times N_s$ and $N_s \times N_w$ matrices, respectively. The evolving factor analysis (EFA), principal component analysis (PCA), or singular value decomposition (SVD) methods can be utilized to determine N_s, prior to the MCR-ALS analysis (Maeder, 1987). Based on the results of the EFA and SVD calculations, we selected $Ns = 3$ for the present analysis of the CD spectral data for the complex of dsDNA and peptide. To generate the initial guess of the spectral factor S, we employed the EFA and the purest variable detection method implemented in the program package MCR-ALS (ver. 2.0) (the latter method was used to generate the initial guess as an alternative to that obtained by the EFA) (Jaumot, de Juan, & Tauler, 2015; Windig & Guilment, 1991; Windig & Stephenson, 1992).

The aim of the MCR-ALS calculation is to extract C and S^T, which are optimally fitted to the experimental data D with a small residual matrix E.

Unlike the PCA, which is characterized by an orthogonal matrix, in the MCR-ALS analysis, the matrices C and S_T are not uniquely determined due to the rotational and intensity ambiguities involved in the theory, and so some restraints are required to determine the C and S^T matrices. For this purpose, the following five constraints (at least) are possible: i) the non-negativity of the concentration, ii) the unimodality of the spectrum, iii) the closure condition (i.e., the sum of the DNA concentrations was equal to 1), iv) one of the spectra should be that of B-DNA, which should be observed in the absence of the peptide, and v) the angles between any two factors should be less than 130 degrees, to hinder the symmetrical properties of the computed spectra. With these optional constraints, C and S^T were iteratively optimized to reproduce the experimental data.

To evaluate the obtained solution, two types of parameter values (lack of fit; *LOF*) are defined, as follows. One is

$$LOF_1 = 100 \times \sqrt{\frac{\sum_{i,j} E_{ij}^{\ 2}}{\sum_{i,j} D_{ij}^{\ 2}}}\ , \tag{7}$$

where D_{ij} and E_{ij} are the (i, j) components of the data matrix D and the residual matrix E, respectively. The other is

$$LOF_2 = 100 \times \sqrt{\frac{\sum_{i,j} e_{ij}^{\ 2}}{\sum_{i,j} d_{ij}^{\ 2}}}\ , \tag{8}$$

where d_{ij} is the (i, j) component of the reproduced data matrix obtained by the PCA, and e_{ij} is defined as

$$e_{ij} = d_{ij} - \left(CS^T\right)_{ij}\ . \tag{9}$$

In the iteration, the relative difference in the standard deviations of the residual matrices between the previous and current steps is defined as follows:

$$R = 100 \times \frac{SD_{PCA}(previous) - SD_{PCA}(current)}{SD_{PCA}(current)}, \tag{10}$$

where

$$SD_{PCA} = \sqrt{\frac{\sum_{i,j} e_{ij}^2}{N_r \times N_w}}, \tag{11}$$

is used as the convergence criterion in the MCR-ALS calculation (Jaumot, Gargallo, de Juan, & Tauler, 2005). This criterion of the iteration was set as |R| = 0.1%, which is the default value of the MCR-ALS program.

In the conventional MCR-ALS results, the LOF_l was 9.24, and its shape contained symmetric properties (Figure 6A). By employing this solution as the initial guess for the MCR-ALS, we performed the second MCR-ALS calculation, and the LOF_l was improved, but the solution still involved similar symmetric properties (Figure 6B). In addition, the symmetric properties became even stronger for the resultant data from the second MCR-ALS calculation, thus indicating that the solutions fell into local minima. In this manner, the overfitting of the solutions occurred in the conventional MCR-ALS calculation. Accordingly, more effective sampling methods must be introduced to solve this problem.

Thus, we extended the conventional MCR-ALS scheme by combining it with the SA method. To the best of our knowledge, our report is the first to propose this type of combination of the MCR-ALS and SA protocols to extend the sampling space (Figure 7). Our scheme is as follows. First, the virtual initial temperature T was set. In our study, we set the initial virtual

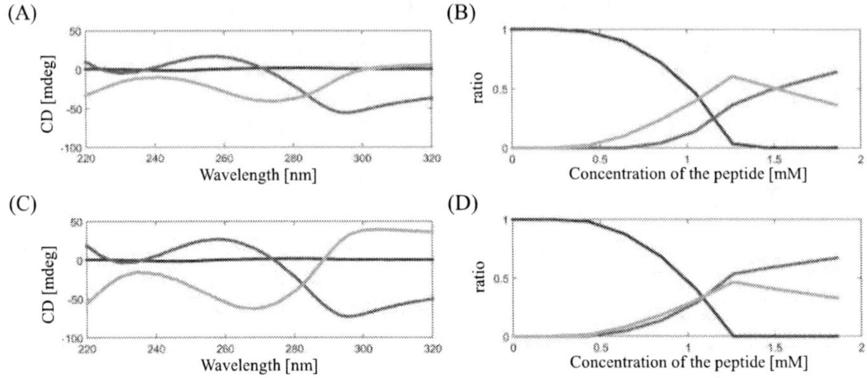

Figure 6. Spectra and concentrations obtained by the conventional MCR-ALS calculations. The spectra and concentrations that were obtained by employing the initial guess derived from the EFA are shown in (A) and (B), respectively (LOF_I = 9.02). By adopting the obtained solution as the initial guess, the MCR-ALS calculation was iteratively performed two times, and the resultant spectra and fractions are depicted in (C) and (D), respectively (LOF_I = 8.09). Symmetrical properties were found in the spectra (A, C). The spectral and fraction profiles that are derived from the three components are depicted in distinct colors, with the profiles of B-DNA, DNA1, and DNA2 colored blue, yellow, and red, respectively. © [2017] Kang, J., Yamasaki, K., Sano, K., Tsutsui, K., Tsutsui, K. M. & Tateno, M.. Originally published in *J. Phys. Soc. Jpn.*, *86*, 014802 (2017).

temperature as T=15. The initial guess of the spectral profile S was generated by the EFA. Second, we performed 500 cycles of the MCR-ALS calculation at the constant temperature (Figure 7; for details, 7B). Third, we saved all of the spectra factors that were converged.

The temperature for the next 500 cycles of the MCR-ALS was then selected, as follows. If the minimum LOF_I value among those obtained in the current 500-cycle iteration was less than that of the previous iteration, then the S^T value corresponding to the former LOF_I and the current temperature minus 5 were employed as the initial guess of the S^T value and the temperature in the subsequent step, respectively (for the other case, see Figure 7A). This was performed until the number of iterations reached the maximum number (100-step iterations were done in our study).

The detailed procedure is described in Figure 7B, with respect to the MCR-ALS stage (constant temperature was imposed during 500 cycles) and the update of S in the SA-MCR-ALS scheme. In every cycle of the MCR-ALS stage, we examined whether the obtained S should be accepted; here, the heat bath method was adopted, as follows: If the difference between the currently- and previously-converged LOF_l values, referred to as Δ, is negative, or if $\exp(-\Delta/T) > \mathrm{Rand}(0,1)$ is satisfied, then the current spectral factor S is accepted. This S matrix is then updated by perturbing the elements by random values that follow the Gaussian distribution function; i.e., $N(\mu = 0, \sigma = 1)$, as follows:

$$S = S - 100 + 200 \times N(0,1).$$

The convergence criterion in the MCR-ALS stage was $|R| < 0.1$ (this is the default value in the MCR-ALS MATLAB toolbox (Jaumot, Gargallo, de Juan, & Tauler, 2005), which is the program employed as the platform for our analysis). In addition, the iterations of the MCR-ALS calculations were stopped when the number of cycles reached the limitation (500 cycles).

As a result of our analysis, we found that the best solution of the SA-MCR-ALS did not exhibit the symmetric properties that were observed in the conventional MCR-ALS. Nevertheless, the LOF_l value of the solution obtained by the SA-MCR-ALS scheme was 8.08, which was better than that of the conventional MCR-ALS (9.24). Thus, the analysis employing the SA-MCR-ALS calculations successfully extracted three types of DNA conformations from our CD spectra, which correspond to the blue, yellow, and red lines; i.e., B-DNA, DNA1, and DNA2, respectively, as shown in Figure 8. In the absence of the peptide, almost all DNA was in the B-DNA state. As the peptide concentration increased, the fractions of DNA1 increased up to 80% at the 1.1 μM peptide concentration, but then decreased to 40% at the 1.9 μM peptide concentration. By contrast, the concentration of DNA2 monotonously increased up to 60%, as the peptide concentration increased (Figure 8).

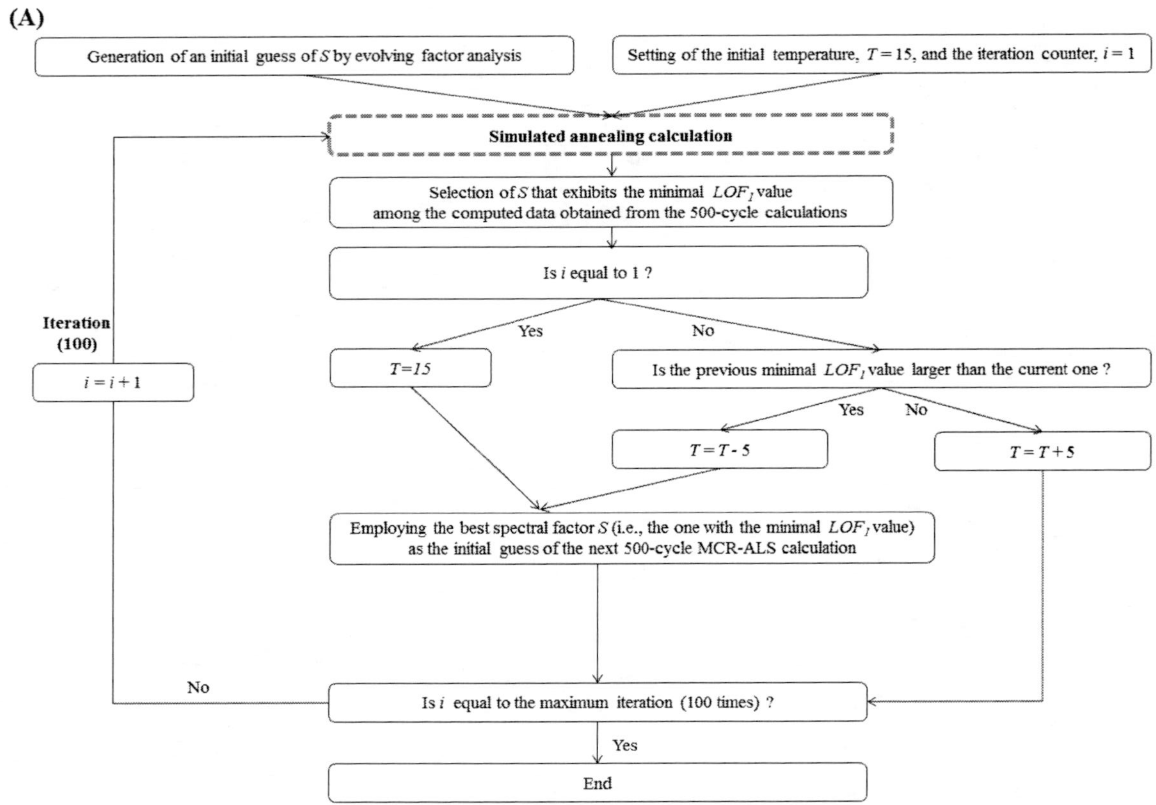

Figure 7. (Continued).

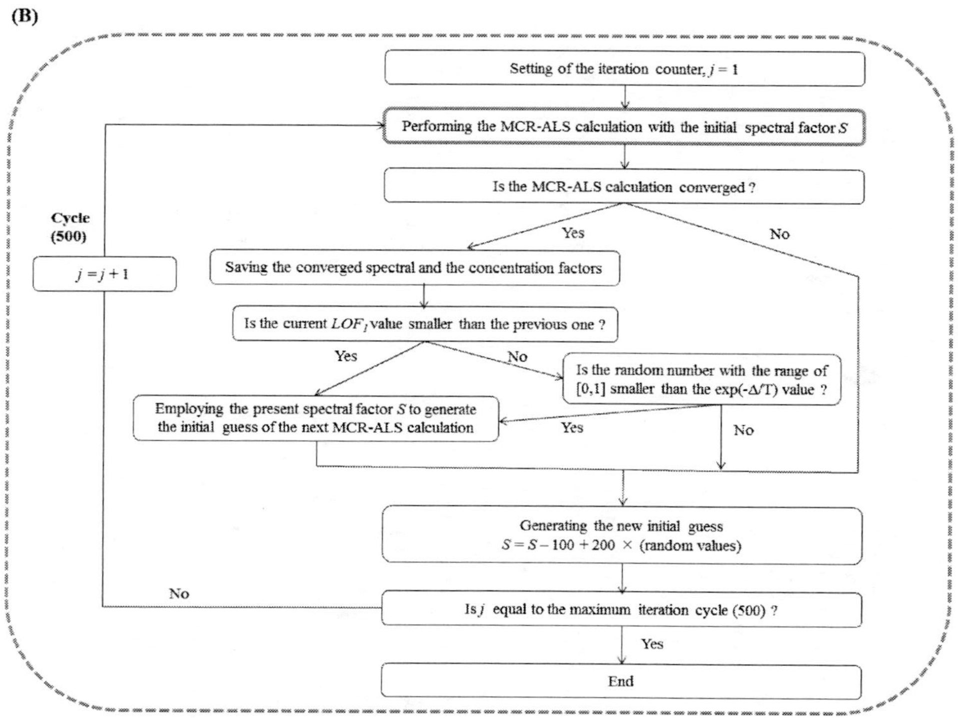

Figure 7. Protocol of our SA-MCR-ALS analysis. (A) The overall procedure is shown with the temperature control algorithm. (B) The detailed procedures for the dashed blue box shown in (A). The green box depicted in (B) represents the conventional MCR-ALS calculation, and Δ represents the difference between the previous and current LOF_1 values. © [2017] Kang, J., Yamasaki, K., Sano, K., Tsutsui, K., Tsutsui, K. M. & Tateno, M.. Originally published in *J. Phys. Soc. Jpn.*, *86*, 014802 (2017).

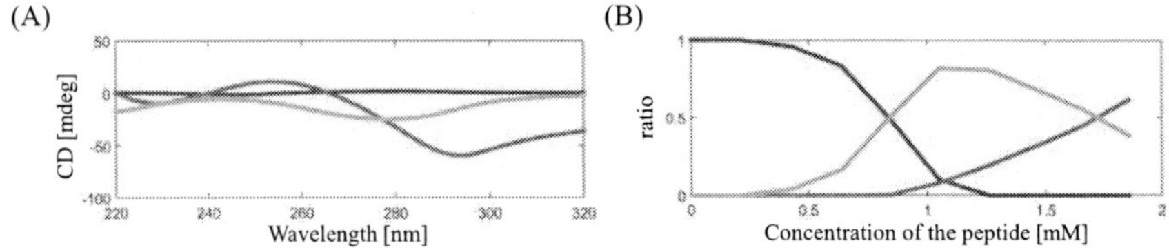

Figure 8. The spectra and concentrations obtained by the SA-MCR-ALS calculations. The spectra and the fractions obtained by our SA-MCR-ALS calculations are shown in (A) and (B), respectively (LOF_1 = 8.08). No symmetrical properties are involved in the spectra (Figure 6C). The spectral and fraction profiles derived from the three components are depicted in distinct colors, with the profiles of B-DNA, DNA1, and DNA2 colored blue, yellow, and red, respectively. © [2017] Kang, J., Yamasaki, K., Sano, K., Tsutsui, K., Tsutsui, K. M. & Tateno, M.. Originally published in *J. Phys. Soc. Jpn.*, *86*, 014802 (2017).

In comparison with the previously reported CD spectra of calf thymus DNA (Cheng & Mohr, 1974; Schwartz & Fasman, 1977), we found that the spectral shape of DNA1 was similar to that of ψ DNA, in which a negative peak with a much larger amplitude than that of the B-DNA spectrum was observed at ~280 nm, corresponding to a peak in the range of 260-280 nm, which is typically observed for ψ DNA (Cheng & Mohr, 1974; Schwartz & Fasman, 1977). Moreover, the CD spectra of DNA2 showed large negative values around 300 nm, which are also a characteristic feature of an extreme condition in the formation of ψ DNA (Schwartz & Fasman, 1977). Thus, although the symmetric properties were not involved, the resultant data exhibited the characteristic features of the ψ DNA structure, supporting the validity of our analysis.

In this manner, our SA-MCR-ALS analysis revealed that the DNA-binding peptide analyzed in this study induced different DNA conformations, which may be similar to ψ DNA. A previous study proposed that the conformation of ψ DNA was close to that of B-DNA (Maniatis, Venable, & Lerman, 1974), although only the restricted structural features of ψ-DNAs are currently known, despite previous intensive investigations. However, our present analysis revealed that ψ DNA may be a mixture of two distinct DNA conformations, DNA1 and DNA2 (defined as the pure spectra depicted in Figure 8), as well as B-DNA, and the relative concentrations depend on the peptide concentration. Namely, more peptide molecules should bind to DNA2 than to DNA1. These observations further suggest that the dynamical processes that form ψ DNA involve at least two distinct steps inducing the different DNA conformations, which are probably similar to DNA1 and DNA2.

In summary, our theoretical scheme is effective to investigate the substantial and detailed properties of multivariate data with significant difficulties in the degrees of freedom, by overcoming the aforementioned problems of the MCR-ALS calculation (i.e., the sampling problem and the initial guess issue). Similar theoretical issues have also been encountered in complicated problems of other fields, such as analyses of time-series data obtained under different conditions, which also suffer from the local minimum problem. Our SA-MCR-ALS analysis is also applicable to those fields, to extend the sampling spaces and to resolve the local minimum problem and the initial guess issue. Thus, the SA method is a fundamental scheme that is extremely applicable to resolve the local minimum problem, in various scientific fields relevant to biological macromolecular systems.

ACKNOWLEDGMENTS

J.K. was supported by the Korea Research Fellowship Program through the National Research Foundation of Korea (NRF) funded by the Ministry of Science and ICT (2017H1D3A1A01053094). H.Y. and M.T. were supported by a Grant-in-Aid from the Ministry of Education, Culture, Sports, Science and Technology of Japan (MEXT) under contract No. 17K10357.

REFERENCES

Anderson, A. C. (2003). The process of structure-based drug design. *Chem. Biol.*, *10* (9), 787-797.

Anderson, J. B. (1976). Quantum chemistry by random walk. H 2P, H+3 D3h 1A′1, H2 3Σ+u, H4 1Σ+g, Be 1S. *J. Chem. Phys.*, *65* (10), 4121-4127.

Andersson, Y., Langreth, D. C. & Lundqvist, B. I. (1996). van der Waals Interactions in Density-Functional Theory. *Phys. Rev. Lett.*, *76* (1), 102-105.

Becke, A. D. (1993). Density-Functional Thermochemistry. 3. The Role of Exact Exchange. *J. Chem. Phys.*, *98* (7), 5648-5652.

Binda, C., Wang, J., Pisani, L., Caccia, C., Carotti, A., Salvati, P. & Mattevi, A. (2007). Structures of human monoamine oxidase B complexes with selective noncovalent inhibitors: safinamide and coumarin analogs. *J. Med. Chem.*, *50* (23), 5848-5852.

Cerny, J. & Hobza, P. (2007). Non-covalent interactions in biomacromolecules. *Phys. Chem. Chem. Phys.*, *9* (39), 5291-5303.

Chan, D. I., Prenner, E. J. & Vogel, H. J. (2006). Tryptophan- and arginine-rich antimicrobial peptides: structures and mechanisms of action. *Biochim. Biophys. Acta*, *1758* (9), 1184-1202.

Cheng, S. M. & Mohr, S. C. (1974). The thermal transition of ‘psi’ DNA monitored by circular dichroism. *FEBS Lett.*, *49* (1), 37-42.

Chien-Cheng, C., Leuo-Hong, W., Cheng-Yan, K., Ming, O. & Wen-Chin, C. (1998). Molecular binding in structure-based drug design: a case study of the population-based annealing genetic algorithms. *Proc. Int. C. Tools Art.*, 328-335.

Dougherty, D. A. (1996). Cation-π interactions in chemistry and biology: a new view of benzene, Phe, Tyr, and Trp. *Science*, *271* (5246), 163-168.

Dougherty, D. A. (2007). Cation-π interactions involving aromatic amino acids. *J. Nutr.*, *137* (6 Suppl 1), 1504S-1508S, discussion 1516S-1517S.

Farver, O., Grell, E., Ludwig, B., Michel, H. & Pecht, I. (2006). Rates and Equilibrium of CuA to Heme a Electron Transfer in Paracoccus denitrificans Cytochrome c Oxidase. *Biophys. J.*, *90* (6), 2131-2137.

Felten, J., Hall, H., Jaumot, J., Tauler, R., de Juan, A. & Gorzsas, A. (2015). Vibrational spectroscopic image analysis of biological material using multivariate curve resolution-alternating least squares (MCR-ALS). *Nat. Protoc.*, *10* (2), 217-240.

Gallivan, J. P. & Dougherty, D. A. (1999). Cation-π interactions in structural biology. *Proc. Natl. Acad. Sci. USA.*, *96* (17), 9459-9464.

Garrido, M., Rius, F. X. & Larrechi, M. S. (2008). Multivariate curve resolution–alternating least squares (MCR-ALS) applied to spectroscopic data from monitoring chemical reactions processes. *Anal. Bioanal. Chem.*, *390* (8), 2059-2066.

Goodsell, D. S., Morris, G. M. & Olson, A. J. (1996). Automated docking of flexible ligands: applications of AutoDock. *J. Mol. Recognit.*, *9* (1), 1-5.

Haeusler, A. R., Donnelly, C. J., Periz, G., Simko, E. A. J., Shaw, P. G., Kim, M. S. & Wang, J. (2014). C9orf72 nucleotide repeat structures initiate molecular cascades of disease. *Nature*, *507* (7491), 195-200.

Hagiwara, Y., Kang, J. & Tateno, M. (2011). Structural Instability of the Active Site of T1 Lipase Induced by Replacement of Na(+) with Water Complexed with the Phenylalanine Aromatic Ring. *J. Chem. Theory Comput.*, *7* (8), 2593-2599.

Hagiwara, Y., Matsumura, H. & Tateno, M. (2009). Functional Roles of a Structural Element Involving Na+-π Interactions in the Catalytic Site of T1 Lipase Revealed by Molecular Dynamics Simulations. *J. Am. Chem. Soc.*, *131* (46), 16697-16705.

Hagiwara, Y., Ohta, T. & Tateno, M. (2009). QM/MM hybrid calculation of biological macromolecules using a new interface program connecting QM and MM engines. *J. Phys. Condens. Mat.*, *21* (6), 064234.

Hagiwara, Y. & Tateno, M. (2009). A novel computational scheme for accurate and efficient evaluation of π–π and π–σ stacking. *J. Phys. Condens. Mat.*, *21* (24), 245103.

Hagiwara, Y. & Tateno, M. (2010). Recent advances in jointed quantum mechanics and molecular mechanics calculations of biological macromolecules: schemes and applications coupled to ab initio calculations. *J. Phys. Condens. Mat.*, *22* (41), 413101.

Head-Gordon, M., Pople, J. A. & Frisch, M. J. (1988). MP2 energy evaluation by direct methods. *Chem. Phys. Lett.*, *153* (6), 503-506.

Hwang, H. J., Berry, S. M., Nilges, M. J. & Lu, Y. (2005). Axial methionine has much less influence on reduction potentials in a CuA center than in a blue copper center. *J. Am. Chem. Soc.*, *127* (20), 7274-7275.

Jaumot, J., de Juan, A. & Tauler, R. (2015). MCR-ALS GUI 2.0: New features and applications. *Chemometr. Intell. Lab.*, *140*, 1-12.

Jaumot, J., Gargallo, R., de Juan, A. & Tauler, R. (2005). A graphical user-friendly interface for MCR-ALS: a new tool for multivariate curve resolution in MATLAB. *Chemometr. Intell. Lab.*, *76* (1), 101-110.

Jaumot, J. & Tauler, R. (2010). MCR-BANDS: A user friendly MATLAB program for the evaluation of rotation ambiguities in Multivariate Curve Resolution. *Chemometr. Intell. Lab.*, *103* (2), 96-107.

Kang, J., Kino, H. & Tateno, M. (2011). A theoretical investigation of the functional role of the axial methionine ligand of the Cu(A) site in cytochrome c oxidase. *Biochim. Biophys. Acta*, *1807* (10), 1314-1327.

Kang, J., Ohta, T., Hagiwara, Y., Nishikawa, K., Yamamoto, T., Nagao, H. & Tateno, M. (2009). Electronic and geometric structures of the blue copper site of azurin investigated by QM/MM hybrid calculations. *J. Phys. Condens. Mat.*, *21* (6), 064235.

Kang, J., Yamasaki, K., Sano, K., Tsutsui, K., Tsutsui, K. M. & Tateno, M. (2017). Simulated Annealing-Extended Sampling for Multicomponent Decomposition of Spectral Data of DNA Complexed with Peptide. *J. Phys. Soc. Jpn.*, *86*, 014802.

Ma, J. C. & Dougherty, D. A. (1997). The Cationminus signpi Interaction. *Chem. Rev.*, *97* (5), 1303-1324.

Macalino, S. J., Gosu, V., Hong, S. & Choi, S. (2015). Role of computer-aided drug design in modern drug discovery. *Arch. Pharm. Res.*, *38* (9), 1686-1701.

Maeder, M. (1987). Evolving factor analysis for the resolution of overlapping chromatographic peaks. *Anal. Chem.*, *59* (3), 527-530.

Maizels, N. (2015). G4-associated human diseases. *EMBO Rep.*, *16* (8), 910-922.

Maniatis, T., Venable, J. H. & Lerman, L. S. (1974). The structure of ψ DNA. *J. Mol. Biol.*, *84* (1), 37-64.

Matsumura, H., Yamamoto, T., Leow, T. C., Mori, T., Salleh, A. B., Basri, M. & Rahman, R. N. Z. R. A. (2008). Novel cation-π interaction revealed by crystal structure of thermoalkalophilic lipase. *Proteins*, *70* (2), 592-598.

Meshki, M., Behpour, M. & Masoum, S. (2015). Application of multivariate curve resolution alternating least squares method for determination of caffeic acid in the presence of catechin interference. *Anal. Biochem.*, *473*, 80-88.

Meyer, E. A., Castellano, R. K. & Diederich, F. (2003). Interactions with Aromatic Rings in Chemical and Biological Recognition. *Angew. Chem. Int. Ed.*, *42* (11), 1210-1250.

Minoux, H. & Chipot, C. (1999). Cation-π interactions in proteins: Can simple models provide an accurate description? *J. Am. Chem. Soc.*, *121* (44), 10366-10372.

Moser, C. C., Page, C. C. & Dutton, P. L. (2006). Darwin at the molecular scale: selection and variance in electron tunnelling proteins including cytochrome c oxidase. *Philos Trans. R Soc. Lond B Biol. Sci.*, *361* (1472), 1295-1305.

Muramoto, K., Hirata, K., Shinzawa-Itoh, K., Yoko-o, S., Yamashita, E., Aoyama, H. & Yoshikawa, S. (2007). A histidine residue acting as a controlling site for dioxygen reduction and proton pumping by cytochrome c oxidase. *Proc. Natl. Acad. Sci. USA.*, *104* (19), 7881-7886.

Olmos, V., Benítez, L., Marro, M., Loza-Alvarez, P., Piña, B., Tauler, R. & de Juan, A. (2017). Relevant aspects of unmixing/resolution analysis for the interpretation of biological vibrational hyperspectral images. *TrAC Trend. Anal. Chem.*, *94*, 130-140.

Ota, N. & Agard, D. A. (2001). Binding mode prediction for a flexible ligand in a flexible pocket using multi-conformation simulated annealing pseudo crystallographic refinement. *J. Mol. Biol.*, *314* (3), 607-617.

Papa, S., Capitanio, N. & Capitanio, G. (2004). A cooperative model for proton pumping in cytochrome c oxidase. *Biochim. Biophys. Acta*, *1655* (1-3), 353-364.

Pople, J. A., Head-Gordon, M. & Raghavachari, K. (1987). Quadratic configuration interaction. A general technique for determining electron correlation energies. *J. Chem. Phys.*, *87* (10), 5968-5975.

Regan, J. J., Ramirez, B. E., Winkler, J. R., Gray, H. B. & Malmstrom, B. G. (1998). Pathways for electron tunneling in cytochrome c oxidase. *J. Bioenerg. Biomembr.*, *30* (1), 35-39.

Robinson, H., Ang, M. C., Gao, Y. G., Hay, M. T., Lu, Y. & Wang, A. H. (1999). Structural basis of electron transfer modulation in the purple CuA center. *Biochemistry*, *38* (18), 5677-5683.

Ruckebusch, C. & Blanchet, L. (2013). Multivariate curve resolution: A review of advanced and tailored applications and challenges. *Anal. Chim. Acta*, *765*, 28-36.

Schwartz, A. M. & Fasman, G. D. (1977). Interaction of DNA with poly(L-Lys-L-Ala-Gly) and poly(L-Lys-L-Ala-L-Pro). Circular dichroism and thermal denaturation studies. *Biochemistry*, *16* (10), 2287-2299.

Sliwoski, G., Kothiwale, S., Meiler, J. & Lowe, E. W. Jr. (2014). Computational methods in drug discovery. *Pharmacol. Rev.*, *66* (1), 334-395.

Tsutsui, K. M., Sano, K., Hosoya, O., Miyamoto, T. & Tsutsui, K. (2011). Nuclear protein LEDGF/p75 recognizes supercoiled DNA by a novel DNA-binding domain. *Nucleic Acids Res.*, *39* (12), 5067-5081.

Tsuzuki, S., Honda, K., Uchimaru, T., Mikami, M. & Tanabe, K. (2002). Origin of Attraction and Directionality of the π/π Interaction: Model Chemistry Calculations of Benzene Dimer Interaction. *J. Am. Chem. Soc.*, *124* (1), 104-112.

Tsuzuki, S., Uchimaru, T. & Mikami, M. (2006). Intermolecular Interaction between Hexafluorobenzene and Benzene: Ab Initio Calculations Including CCSD(T) Level Electron Correlation Correction. *J. Phys. Chem. A*, *110* (5), 2027-2033.

Wang, K., Geren, L., Zhen, Y., Ma, L., Ferguson-Miller, S., Durham, B. & Millett, F. (2002). Mutants of the CuA site in cytochrome c oxidase of Rhodobacter sphaeroides: II. Rapid kinetic analysis of electron transfer. *Biochemistry*, *41* (7), 2298-2304.

Wentzell, P. D., Karakach, T. K., Roy, S., Martinez, M. J., Allen, C. P. & Werner-Washburne, M. (2006). Multivariate curve resolution of time course microarray data. *BMC Bioinformatics*, *7*, 343.

Windig, W. & Guilment, J. (1991). Interactive self-modeling mixture analysis. *Anal. Chem.*, *63* (14), 1425-1432.

Windig, W. & Stephenson, D. A. (1992). Self-modeling mixture analysis of second-derivative near-infrared spectral data using the SIMPLISMA approach. *Anal. Chem.*, *64* (22), 2735-2742.

Wolffe, A. P. (2001). Chromatin remodeling: why it is important in cancer. *Oncogene*, *20* (24), 2988-2990.

Zhen, Y., Schmidt, B., Kang, U. G., Antholine, W. & Ferguson-Miller, S. (2002). Mutants of the CuA Site in Cytochrome c Oxidase of Rhodobacter sphaeroides: I. Spectral and Functional Properties†. *Biochemistry*, *41* (7), 2288-2297.

Biographical Sketches

Prof. Jiyoung Kang

Affiliation: Institute of Human Complexity and Systems Science, System Science Center for Brain and Cognition, Yonsei University, Seoul, Republic of Korea

Education: Graduated from Department of Physics, University of Tsukuba

Publications from the Last 3 Years:

Kang, J., Pae, C., & Park, H. (2017). Energy landscape analysis of the subcortical brain network unravels system properties beneath resting state dynamics, *Neuroimage*, *149*, 153-164.

Kang, J., Yamasaki, K., Sano, K., Tsutsui, K., Tsutsui, K. M. & Tateno, M. (2017). Simulated Annealing-Extended Sampling for Multicomponent Decomposition of Spectral Data of DNA Complexed with Peptide. *J. Phys. Soc. Jpn.*, *86*, 014802.

Kang, J., Kino, H., Field, J. M., & Tateno, M. (2017). Electronic Structure Rearrangements in Hybrid Ribozyme/Protein Catalysis. *J. Phys. Soc. Jpn.*, *86*, 044801.

Kang, J., & Tateno, M. (2017). Effects of a Positive Feedback Loop upon the Stability of Bi-Connected Elementary Biochemical Reaction Cycles. *J. Phys. Soc. Jpn.*, *86*, 074802.

Sugimoto, A., Sumi, T., Kang, J., and Tateno, M. (2017). Novel Strategy for Discrimination of Transcription Factor Binding Motifs Employing Mathematical Neural Network. *J. Phys. Soc. Jpn.*, *86*, 074803.

Shimada, S., Shinzawa−Itoh, K., Baba, J., Aoe, S., Shimada, A., Yamashita, E., Kang, J., Tateno, M., Yoshikawa, S., & Tsukihara, T. (2017). Complex structure of cytochrome c–cytochrome c oxidase reveals a novel protein–protein interaction mode, *EMBO J.*, *86*, 291-300.

Kim, J., Kang, J., Nishigami, H., Kino, H., & Tateno, M. (2018). Ab Initio Electronic Structure Calculation of [4Fe-3S] Cluster of Hydrogenase as Dihydrogen Dissociation/Production Catalyst. *J. Phys. Soc. Jpn.*, *87*, 034804.

Prof. Jiyoung Kang, is a theoretical biophysicist, graduated from Department of Physics, University of Tsukuba (the Tateno laboratory). She worked for the University of Hyogo. She has been working on developments and biological applications of hybrid *ab initio* QM/MM MD calculations to elucidate catalytic mechanisms of various biological huge catalysts. She is working to integrate physics, structure-based computations, and bioinformatics and mathematical/statistical techniques, toward understanding of higher-order biological functions (such as memory and learning in brain).

Takuya Sumi

Affiliation: Graduate School of Life Science, University of Hyogo, 3-2-1, Kouto, Kamigori-cho, Japan

Education: PhD student of the Graduate School of Life Science, University of Hyogo

Takuya Sumi is a PhD student of the Theoretical and Computational Biology, Graduate School of Life Science. He studies computer simulations in molecular design fields, such as protein-ligand docking calculations, and also contributes to *ab initio* identification of transcription factor binding motifs and its fundamental theory based on mathematical manifold. He is supervised by the other three authors.

Prof. Hiroshi Yamaguchi

Affiliation: Nagoya University Brain & Mind Research Center, 65 Tsurumai-cho, Syowa-ku, Nagoya, Japan

Education: Graduated from Graduate School of Engineering, Tottori University

Prof. Hiroshi Yamaguchi, is an experimental chemist (his current principal field is radiochemistry), graduated from Graduate School of Engineering, Tottori University. He worked for National Institute of Advanced Sciences and Technologies (AIST), National Institute for Longevity Sciences (NILS), Akita Research Institute for Brain and Blood Vessels, and Nagoya University (present). Currently, he is designated lecturer of Nagoya University Brain & Mind Research Center. In recent year, he focused on the development of PET reagents for visualizing glutamate and dopamine neurotransmission systems, and the synthesis of various PET agents targeting tau and amyloid as a clinical study.

Prof. Masaru Tateno

Affiliation: Graduate School of Life Science, University of Hyogo, 3-2-1, Kouto, Kamigori-cho, Japan

Education: Graduated from Department of Biophysics and Biochemistry, Faculty of Science, University of Tokyo

Publications from the Last 3 Years: See above (under Kang).

Prof. Masaru Tateno, male, theoretical biophysicist, graduated from Department of Biophysics and Biochemistry, Faculty of Science, University of Tokyo. He worked for RIKEN, National Institute of Advanced Sciences and Technologies (AIST), Tokyo Institute of Technology (TIT), University of Tsukuba, and University of Hyogo (present). Currently, he is professor of the Division of the Theoretical and Computational Biology in University of Hyogo. In recent year, he focused on developments and biological applications of i) hybrid *ab initio* molecular dynamics calculation system to elucidate functional mechanisms, and ii) statistical pattern recognition theory and techniques toward understandings of biological functions based on genome and mathematical sciences.

In: Simulated Annealing
Editors: Alex Scollen et al.
ISBN: 978-1-53613-674-6

Chapter 3

PHOTOVOLTAIC SYSTEM MAXIMUM POWER EXTRACTION: AN APPLICATION OF SIMULATED ANNEALING

Sarah Lyden[1,*] and Md Enamul Haque[2]
[1]School of Engineering, University of Tasmania, Hobart, Australia
[2]School of Engineering, Deakin University, Geelong, Australia

ABSTRACT

Photovoltaic (PV) Systems are becoming increasingly common, particularly at residential scale, as society becomes more conscious of the environmental impact of conventional electricity generation. PV systems are made up of a number of PV cells joined together in series and parallel arrangements, which are then connected to a local load or utility grid through power electronics interfaces such as DC-DC converters and DC-AC inverters. These power electronics interfaces are necessary to enable the PV system to be controlled and to mitigate the impact of non-linear power-voltage (P-V) characteristics that arise in PV systems. In particular, in a residential environment, a PV system may be partially shaded by a tree or other structure which leads to a multi-peak P-V characteristic.

[*] Corresponding author: Email: Sarah.Lyden@utas.edu.au.

Maximum Power Point Tracking (MPPT) is usually implemented in the DC-DC converter to ensure that the maximum power can be extracted from a PV system under these conditions. The area of MPPT is quite broad and many techniques have been adopted ranging from conventional techniques which can track a MPP under uniform conditions, to those that are designed specifically for the multi-peak scenario. The global searching capability of the Simulated Annealing (SA) technique makes it a good candidate method for solving the PV system MPPT problem in residential environments even with quite arbitrarily selected parameters. This chapter will detail the development of the SA based MPPT method and highlight simulations evaluating the performance of the technique. Special treatment will be given to the parameters of the SA method and which of these have most influence based on the PV system configuration.

Keywords: photovoltaic, simulated annealing, maximum power point tracking

INTRODUCTION

As the world becomes more aware of impending challenges such as global warming, energy shortages, and the number of people who still do not have access to electricity, the development of new and sustainable energy resources is increasing. Solar energy has great potential, as it is available all around the world. In fact, data shows that the incident amount of sunlight on the Earth's surface exceeds 7500 times the world's current total annual primary energy consumption (World Energy Council, 2013). As technology improves, the potential of solar energy resources become more significant.

Typically a photovoltaic (PV) system is made up of a number of PV cells connected in series and parallel to form modules. Multiple modules are then connected in series and parallel arrangements to meet system design requirements, similar to the system shown in Figure 1. One of the limitations of this system is that PV cells have a non-linear current-voltage (I-V) and power-voltage (P-V) characteristic as shown in Figure 2. In Figure 2 it can be seen that there is a unique point on the P-V characteristic which corresponds to maximum available power for a given set of environmental conditions. This unique point will change with the environmental conditions and is particularly susceptible to irradiance, cell temperature and cell shading. When multiple PV modules are connected together to form a system, the non-linearity of the P-V characteristic becomes more severe due to the presence of bypass diodes. These diodes, as shown across the modules in Figure 1, are used to ensure that the cell or module with the lowest irradiance does not constrain the operation of the entire system, but does leads to increased complexity in the P-V characteristics as shown in Figure 3. In the case of Figure 3, Maximum Power Point Tracking (MPPT) becomes a global optimisation problem requiring continuous searching due to potentially rapidly changing environmental conditions.

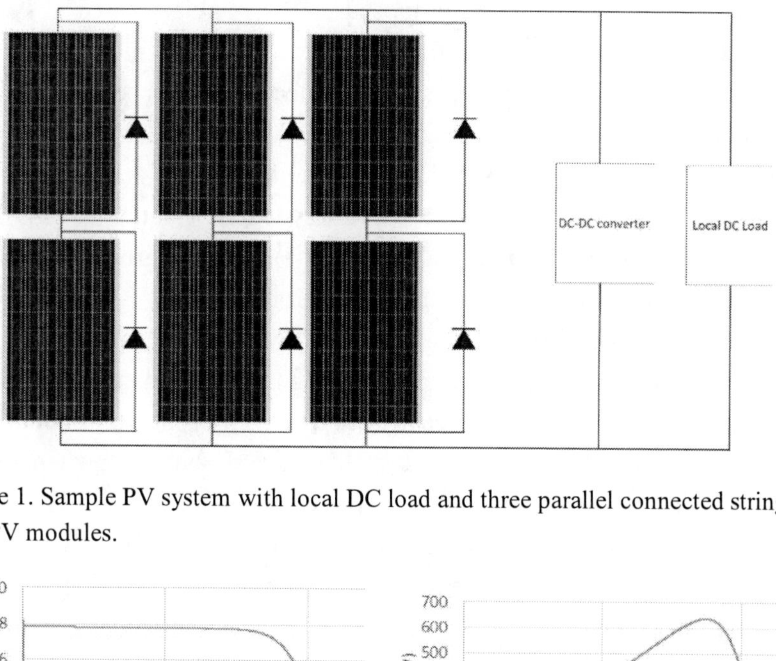

Figure 1. Sample PV system with local DC load and three parallel connected strings of two PV modules.

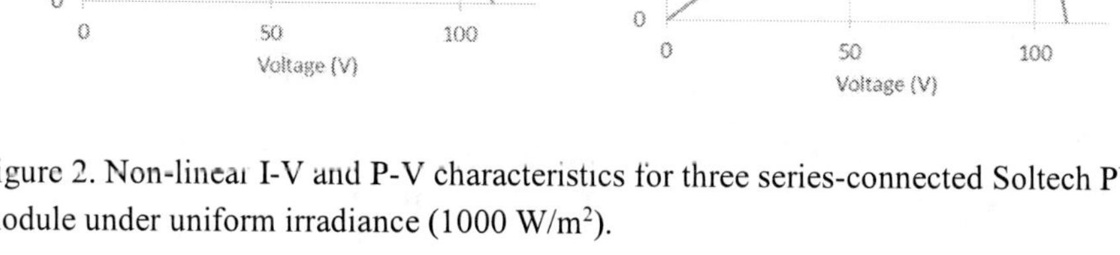

Figure 2. Non-linear I-V and P-V characteristics for three series-connected Soltech PV module under uniform irradiance (1000 W/m^2).

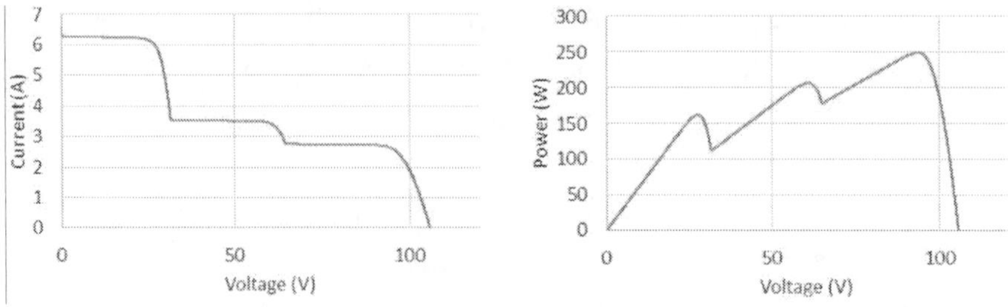

Figure 3. Increased non-linearity in P-V characteristics under partial shading for three series connected Soltech PV modules (irradiance 800 W/m^2, 450 W/m^2, 350 W/m^2).

MPPT is a very broad area of research for PV system optimisation. Multiple techniques have been proposed as the popularity of PV systems increases. These techniques vary from simple techniques which approximate the MPP location, to advanced searching techniques based on artificial intelligence and key properties of the P-V characteristic. Simulated Annealing (SA) has been explored as a suitable GMPPT technique by the authors of this chapter (prior work see (Lyden & Haque, 2015a, 2014, 2015b, 2016a; Lyden, Haque, & Xiao, 2014)). SA has great potential for solving the non-linear global optimisation problem for PV systems. Some key properties of the likelihood of the GMPP moving, as also researched by the chapter authors, is a key input to why utilising a time-invariant technique like SA can work for the dynamically changing environmental conditions. Prior work by the authors (Lyden, 2015), has shown that when simple shading classifications are considered, the type of shading conditions which lead to a change in the approximate location of the GMPP with respect to voltage is much slower than the impact on the power output due to rapidly changing irradiance conditions. Based on these characteristics, it is possible for the SA method to locate the GMPP with arbitrarily selected parameters as shown in (Lyden & Haque, 2015a), and then remain around this point by tracking with a local searching MPPT method such as Perturb and Observe (P&O) (Kobayashi, Takano, & Sawada, 2003; Lyden & Haque, 2015b; Sera, Mathe, Kerekes, Spataru, & Teodorescu, 2013) until a significant change in conditions occurs. A global search could then be triggered to ensure that the method consistently searches for the global peak.

Prior work on the SA method for PV MPPT applications has utilised rather arbitrarily selected parameters, and the impact of these parameters has been explored fully for the BP380 PV modules ("BP 380 80 W photovoltaic module datasheet," n.d.) in (Lyden & Haque, 2016a). In this chapter, links between the parameters of the PV system and the performance of the SA based MPPT technique are made. Based on these links, recommendations on the parameters requiring tuning for a particular PV system can be determined. With further studies, a framework for selecting optimal parameters for the SA based MPPT method for PV applications could be established based on the parameters of the system.

SA Based Method of MPPT

SA is based on the process of annealing in metallurgy (Ao & Gelman, 2012). Essentially, when applied to global optimisation problems, SA acts like a guided random search where initially a large variation in the performance is permitted to enable the space to be searched fully and then as the search progresses less variation is permitted. Controlling this search is an artificial temperature parameter which will decrease as the search runs. Initially, prospective new operating points will be compared with the current best operating point to assess their 'fitness'. If this new operating point has a larger power than the current best operating point it will be accepted by default. The interesting part of the search is when the new operating point has a lower power than the current best operating point. In this case, the new operating point could still be accepted as the new best point based on the acceptance probability (Poole & Mackworth, 2010; Zäpfel, Braune, & Bögl, 2010). Essentially, during the early part of the search when the temperature is high, it is more likely that an operating point with a lower power will be accepted, and this enables the method to search more fully without becoming trapped in a region of a local maxima. The acceptance probability is given in (1) (Černý, 1985; Nourani & Andresen, 1998).

$$Pr = \exp\left(\frac{P(k)-P(k-1)}{T(k)}\right) \tag{1}$$

where, $P(k)$ is the power of the sampled operating point, $P(k-1)$ is the current best operating point power, and $T(k)$ is the current temperature of the search.

The temperature parameter will reduce over the searching time due to a cooling schedule. This cooling schedule is one of the key parameters which can influence the performance of the algorithm. In other applications (Chen, Lee, & Wu, 2005; Katsigiannis, Georgilakis, & Karapidakis, 2012; Sutthibun & Bhasaputra, 2010), the SA algorithm is utilised in a time-invariant large scale global optimisation problem, where multiple iterations can be taken to find the global optimal solution to the problem. In a PV system MPPT application the incident irradiance on the system may change rapidly as cloud cover changes throughout the day. An example of how this irradiance changes across a single day under cloud conditions on a scale of one minute variations utilizing data from the Bureau of Meteorology in Australia (Bureau of Meteorology, 2012) is shown in Figure 4. In addition to this variation in irradiance the path of the shadow of objects across the PV system will influence the irradiance which individual cells and modules receive.

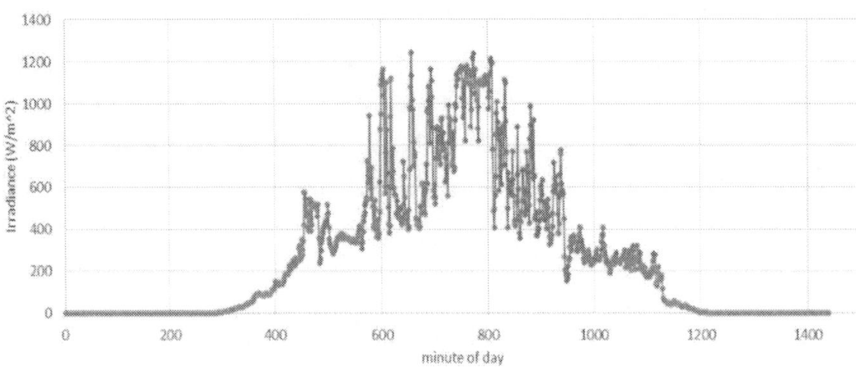

Figure 4. Irradiance variation for a single day in January 2017 at the Cape Grim weather station (Bureau of Meteorology, 2012).

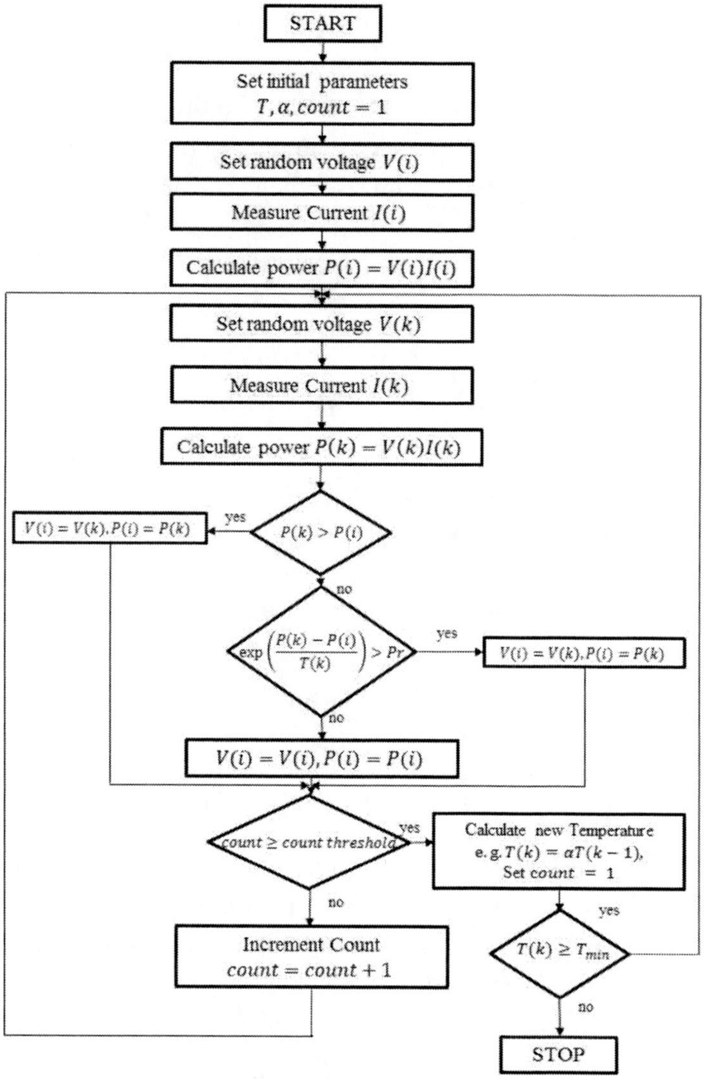

Figure 5. Flowchart of the SA based MPPT method (adapted from (Lyden & Haque, 2016b)).

For application in PV systems, the search time permitted is time-bound as extended searching may encompass time where the conditions have changed and will lead to potentially lower overall energy capture if excessive points at a lower power are sampled. Our prior work has shown that the SA method can locate the neighbourhood of the GMPP with sufficient speed, which means that fine-searching for the exact MPP which will be changing with time is not always necessary. In this sense, the SA based MPPT method is designed to locate the region of the GMPP and then can be coupled with a second stage of searching to find the exact GMPP, by using a conventional MPPT technique such as P&O (Lyden & Haque, 2015b; Sera et al., 2013). The flowchart of the SA based MPPT technique is shown in Figure 5.

SA Parameters

There are a number of factors which could affect the performance of SA based GMPPT methods, including:

- Cooling function – including initial temperature, stopping temperature, cooling rate and frequency
- Threshold for the acceptance probability
- Neighbourhood size
- Stopping criterion based on number of cases without improvement

The impact of variations in these parameters on the performance of the SA based GMPPT method for the BP380 PV modules has been considered in (Lyden & Haque, 2016a). These results have shown that the following parameters have the most significant impact on the performance of the SA based GMPPT method when the goal is to locate the region of the GMPP:

- Variable acceptance probability threshold
- Linear neighbourhood reduction function
- Cooling schedule including the superior performance of Lundy cooling schedule and Simulated Quenching (SQ) when compared with the standard geometric cooling schedule.
- Applying a stopping criterion later in the search

Of these possibilities for improving performance, the cooling schedule parameters and neighbourhood size are considered in this chapter. These parameters are considered in the context of the PV system parameters. This study could be extended into a framework for evaluating suitable parameters to apply the SA based GMPPT method to any PV system.

The two cooling schedules investigated in this chapter are the Lundy cooling schedule (Cohn & Fielding, 1999), given in (2) and SQ (Coddington, n.d.) given in (3).

$$T(k) = \frac{T(k-1)}{1+\beta T(k-1)} \tag{2}$$

where β is the cooling schedule constant and $T(k-1)$ is the previous temperature.

$$T(k) = \frac{T(0)}{exp(1-c_{sq})k} \tag{3}$$

where $T(0)$ is initial temperature, c_{sq} is the SQ cooling constant, and k is the iteration number.

The literature has shown that the chosen neighbourhood size for searching can have a substantial impact on the performance of the SA algorithm (Ao & Gelman, 2012; Chong & Zak, 2011; Soliman & Mantawy, 2011). To ensure good performance the neighbourhood size should be small enough that the algorithm can converge in a short time frame, but it must also be sufficiently large to present a wide solution space and allow the method to escape from local maxima. Different neighbourhood sizes based on the system configuration are considered in this chapter in addition to a neighbourhood reduction function. In each of these implementations, regardless of whether the neighbourhood size is changing, the current neighbourhood is always defined (without exceeding specified limits) to be centred around the current best operating point. A maximum and minimum voltage range for all implementations considered is defined based on the open-circuit voltage characteristics of the given set of modules, where V_{min} is 60% of the open-circuit voltage of one module and V_{max} is 90% of the open-circuit voltage of the string of modules connected together as given in (4) and (5). This is consistent with observations of PV characteristics showing that the voltage at the MPP is usually around 80% of the open-circuit voltage (Patel & Agarwal, 2008). As the open-circuit voltage will decrease with an increase in cell temperature and decrease in irradiance, a broader range is considered with the manufacturer specifications of per module open-circuit voltage.

$$V_{min} = 0.6V_{oc} \tag{4}$$

$$V_{max} = 0.9N_sV_{oc} \tag{5}$$

Each PV module comes with a manufacturer datasheet which conveys some of the key parameters of the module. These parameters include the open-circuit voltage (V_{oc}), short-circuit current (I_{sc}), and voltage and current at the MPP under standard test conditions (STC) (V_{mpp}, I_{mpp}). In addition to these parameters that apply to the individual module, there are parameters which apply to the entire system. The key parameters here are the number of modules in series and the number of strings in parallel. In this chapter, links are drawn between the parameters utlised in the SA based MPPT method implementation and features of the PV system configuration. By approaching the method in this way, suggestions on the parameters of interest for future implementations are generated creating a list of parameters most essential to tune.

The parameters of the SA based MPPT method explored vary depending on the cooling function implemented. As the Lundy cooling schedule and SQ have been shown to have the best performance in preliminary studies, they will be explored in detail in this chapter. For the Lundy cooling function the main parameter of interest is the cooling rate β. Additional parameters which could be varied to control the search are $T(0), and\ T(final)$. For the SQ implementation the main parameter of interest is the cooling constant c_{sq}. Additional parameters of interest which are not fully explored in this paper are the start and stopping temperature $T(0), T(final)$. With respect to the neighbourhood size, various neighbourhood sizes will be considered along with a dynamically reducing neighbourhood size.

The PV system parameters and SA based MPPT method parameters of interest are summarised in Table 1 below.

Table 1. Parameters considered in the study

PV system parameters	SA based MPPT method parameters
V_{oc}	Lundy β
I_{sc}	SQ c_{sq}
V_{mpp}	Neighbourhood reduction function
I_{mpp}	Initial size
N_s	Reduction rate
N_p	

SIMULATION MODEL

To evaluate the performance of the SA based MPPT technique with different PV system configurations and different parameter values a simple assessment method has been developed. This method involves utilising a simple Matlab/Simulink based simulation model combining PV module models with a voltage source to sweep the characteristics and obtain I-V and P-V characteristics for the configuration. This equivalent model for three-series connected modules is shown in Figure 6. A selection of characteristics featuring uniform conditions, partial shading conditions with MPPs in different locations and characteristics showing minimal differences between irradiance on each module are developed with this model. The condition is specified by the irradiance defined for each module in the system, for instance 1000, 1000, 1000 would represent uniform conditions for a three module system while 1000, 800, 700, 500, 200 would represent a non-uniform condition for the five module system. The same irradiance conditions for each number of series connected modules are applied to the different manufacturer's panels. Multiple iterations (10) are then applied to each characteristic with different parameter values developed for different configurations to ensure that consistent results are obtained. A flowchart of this process is shown in Figure 7. The manufacturer parameters of the selected modules are given in Table 2. The following system configurations are considered:

- Three series connected Soltech 215 W modules
- Three series connected Green Technology 175 W modules
- Three series connected Bosch Solar 240 W modules
- Three series connected BP Solar 175 W modules
- Three series connected Trina Solar 175 W modules
- Five series connected Bosch Solar 240 W modules
- Five series connected BP Solar 175 W modules
- Eight series connected Bosch Solar 240 W modules
- Eight series connected BP Solar 175 W modules
- Twelve series connected Bosch Solar 240 W modules

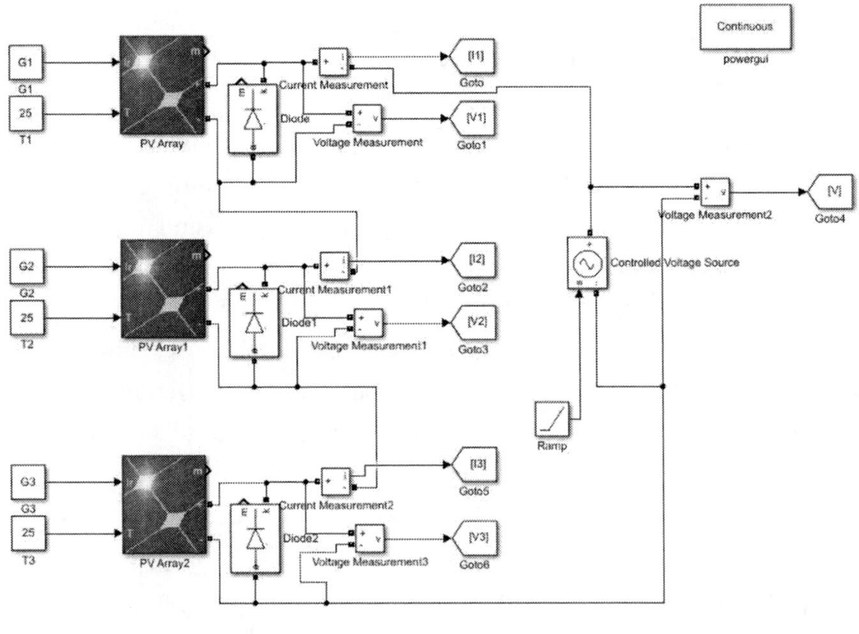

Figure 6. Example simulation model for obtaining the characteristics based on three series connected PV modules.

Table 2. Module parameters for different manufacturers

	1Soltech 1STH-215P	A10Green Technology A10J-S72-175	Bosch Solar Energy c-Si M 60 NA30117 240 W	BP Solar BP4175T	Trina Solar TSM-175 DAO1
I_{sc}	7.84	5.17	8.6	5.45	5.3
V_{oc}	36.3	43.99	37.4	43.6	43.9
V_{mpp}	29	36.63	30	35.4	36.2
I_{mpp}	7.35	4.78	8.1	4.94	4.85
N_s	60	72	60	72	72
P_{mpp}	213.15	175.0914	243	174.876	175.57

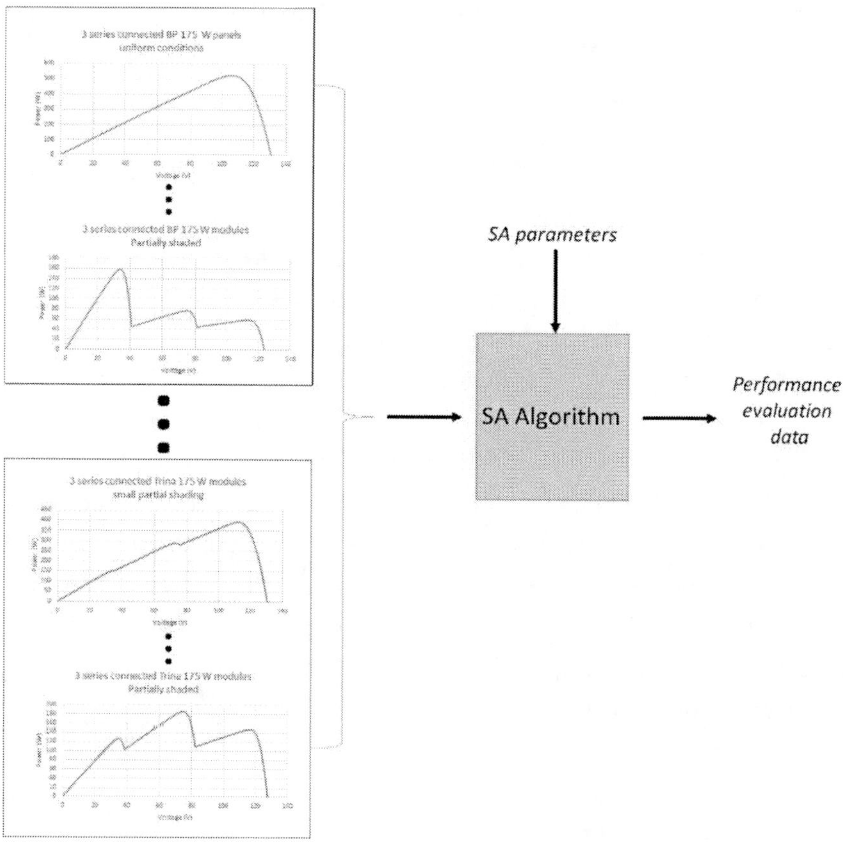

Figure 7. Evaluation model for each SA based MPPT implementation.

Results

The simulation model has been applied with a wide variety of PV module specifications and system specifications to identify if there are suitable trends which can optimise the performance of the system by linking the parameters of the PV system as given in Table 1, with the optimal parameters of the SA based MPPT method implementations. These results are detailed in this section. Two measures of convergence to the GMPP are considered. The first is a 5% convergence rate which assesses if the final operating point is within 5% of the known GMPP power and within 5V of the known GMPP voltage. A tighter tolerance of 2% and 2V is also considered.

Lundy Cooling Schedule

The Lundy cooling schedule is applied to the characteristics for each system configuration with a variety of Lundy cooling constant values ranging from 0.01 to 0.05. As the value of the constant increases the number of samples required in the search decreases, based on a starting temperature of 25°C and final temperature of 0.2°C. With the 5%, 5V tolerance criterion the method converged for each three module system with each set of parameter values 100% of the time. When the tighter 2%, 2V tolerance is considered for the three module systems the method still converges consistently in greater than 97% of cases. This is shown in Figure 8 for the 2%, 2V tolerance. When larger systems are considered consisting of five to 12 modules, the percentage of cases which converge with each Lundy cooling constant reduces slightly. Table 3 shows the average number of cases that converge for each system configuration considered with each set of parameter values for a 5%, 5V tolerance. There is a slight trend that as the number of modules in series increases the likelihood of the method converging to within 5% of the known GMPP power and 5V of the known GMPP voltage decreases for the same Lundy cooling constant. This suggests that for a larger system a smaller value of the Lundy cooling constant should be selected, leading to a larger number of samples in the search space.

Clearly the performance is quite consistent with the different parameter values for the different systems considered, however as the number of samples reduces the average convergence rate does reduce slightly. In MPPT applications it is important to minimise the number of samples required as the environmental conditions may change rapidly. These results suggest that Lundy cooling could be used consistently across different manufacturer modules with arbitrarily selected parameters based on minimising the number of samples required.

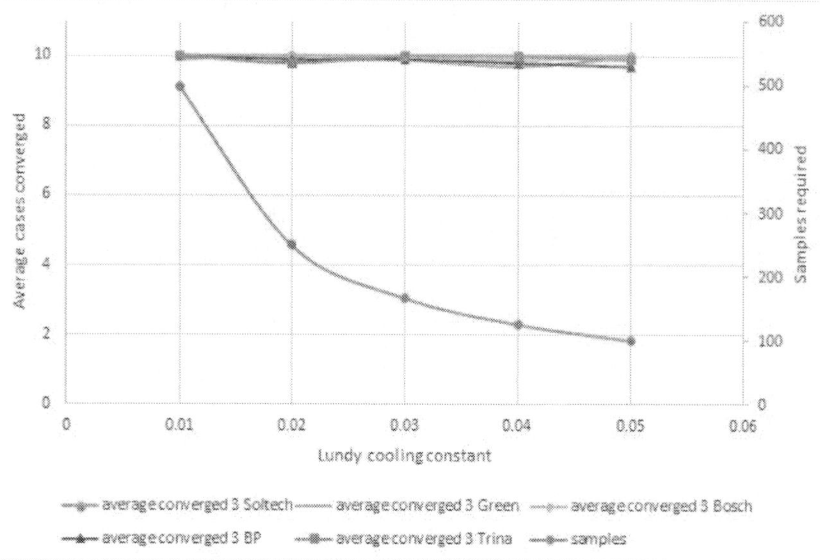

Figure 8. Lundy results for three module configurations.

Simulated Quenching

SQ is considered with cooling constant values ranging from 0.2 to 0.8. These values result in a substantially reduced number of samples being used in the search when compared with the Lundy implementations described

Table 3. Average number of cases that converge with the Lundy cooling schedule for different system configurations

β	samples	average number of cases that converge for each configuration									
		Soltech 3	Green 3	Bosch 3	BP 3	Trina 3	Bosch 5	BP 5	Bosch 8	BP 8	Bosch 12
0.01	497	10	10	10	10	10	10	10	10	9.8	10
0.02	249	10	10	10	10	10	10	10	10	9.5	9.9
0.03	166	10	10	10	10	10	10	10	9.9	9.5	9.8
0.04	125	10	10	10	10	10	10	10	9.6	9.8	9.9
0.05	100	10	10	10	10	10	9.9	10	9.9	9.7	9.4

Table 4. Average number of cases that converge with the Simulated Quenching cooling schedule for different system configurations

c_{sq}	samples	average number of cases that converge for each configuration									
		Soltech 3	Green 3	Bosch 3	BP 3	Trina 3	Bosch 5	BP 5	Bosch 8	BP 8	Bosch 12
0.2	40	9.4	8.8	9.5	9.5	8.8	9.2	8.1	7	7.4	6.7
0.4	50	9.7	9.7	9.5	9.8	9.8	9	9.1	8.3	6.9	6.9
0.6	70	9.9	9.5	9.9	10	9.7	10	9.7	8.7	8.5	8.1
0.8	130	10	9.9	10	10	9.9	10	9.9	9.6	8.9	9.3

above. The starting temperature of 25°C and final temperature of 0.2°C were also used in this implementation. When the 5%, 5V threshold is considered, each three module implementation converges in greater than 94.75% of cases. When the tolerance is tightened to 2%, 2V, the average convergence for the three module implementations drops to greater than 74.5%. There is a general trend of the average convergence increasing as the number of samples increases. Table 4 shows the average number of cases that converge for each system configuration considered with each set of parameter values for a 5%, 5V tolerance. General trends that can be observed are that as the number of samples increases the average convergence also increases across all configurations. It was also generally observed that there were lower average convergence rates for the systems with more modules suggesting that additional samples or more targeted searching may be required within larger systems. When the average convergence rate for each implementation is considered for correlation with the range of open-circuit voltage values, and short-circuit current values on the manufacturer datasheet there is no clear correlation. This suggests that the method works well regardless of the parameter values chosen and that more extensive parameter studies may be required. Figure 9 shows the 2%, 2V tolerance for the SQ implementation.

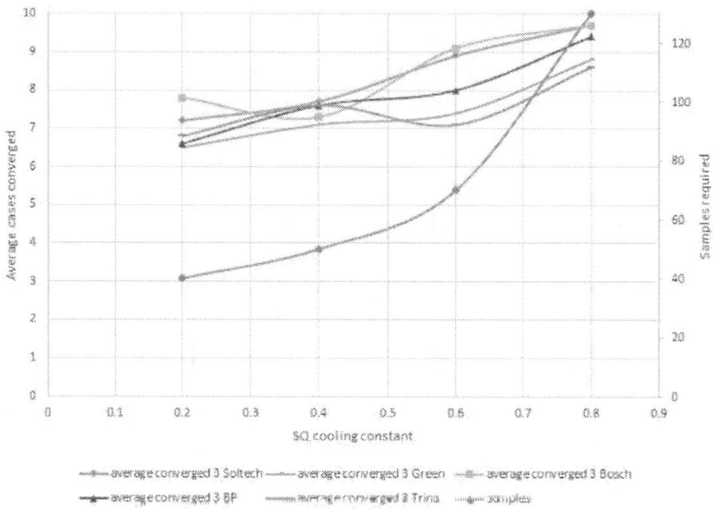

Figure 9. SQ results for three module configurations.

Neighbourhood Reduction Function

The initial neighbourhood size is defined based on the maximum and minimum voltage values defined in (4) – (5). When the temperature is reduced according to the geometric cooling function given in (6), the neighbourhood size is reduced.

$$T(k) = \alpha T(k-1) \tag{6}$$

where, α is the geometric cooling constant, and a value of 0.8 is applied in this study.

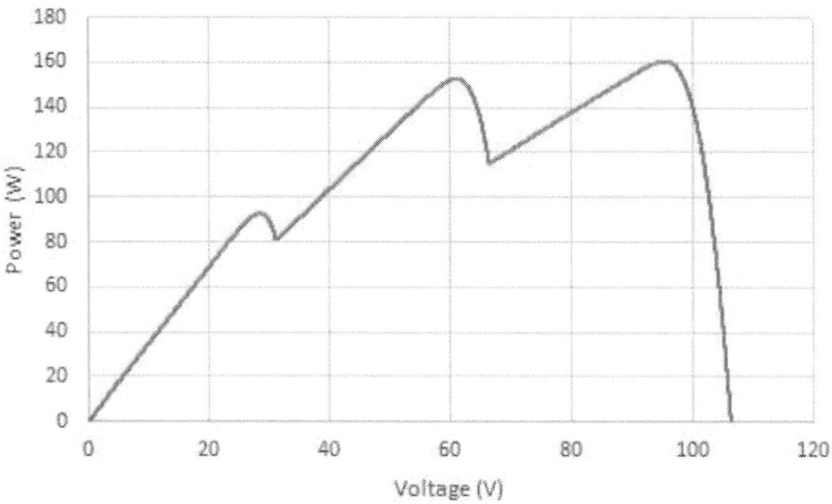

Figure 10. P-V characteristics for conditions leading to similar MPPs.

In the results detailed in this section a linear neighbourhood reduction size is used with a rate of reduction of -0.8125. This relates to the neighbourhood ranges reducing from the initial value defined by (4) – (5), to approximately the value of the open circuit voltage of one module at the end of the search. When the 5%, 5V tolerance criterion is used the method shows good convergence (100%) across most of the three module implementations, but consistently has a low convergence rate for one of the irradiance conditions considered. The particular conditions where the convergence rate is lower relates to a case where there are two peaks of similar value. Sample characteristics for the Bosch 3 series connected modules under these conditions are shown in Figure 10. In these cases, the method may locate the similar value peak as the power values are not significantly different.

With each configuration the method has good overall average convergence, typically converging in greater than 91% of cases. The five module implementations observed the highest convergence rating (100% of cases considered), suggesting that the reduction rate could be tuned to lead to good performance for different system configurations.

Neighbourhood Size

As systems of different configurations are considered in this chapter, it may be necessary to consider a different neighbourhood size for each implementation. Varying neighbourhood sizes are considered ranging from the open-circuit voltage of one module as the range to the open-circuit voltage of the entire system in increments of module open-circuit voltage. In each case the initial sample point is defined in the middle of the voltage range of the panel. These results can show that a large neighbourhood is not necessary to search effectively. The neighbourhood will move dynamically to be centred about the current best operating point enabling a small neighbourhood size to still search a larger space. As previously described, MPPs are known to be separated by about $0.8V_{OC}$, so having the module open-circuit voltage as the searching space should allow a sufficient large searching space that the method can locate an adjacent peak. If additional information is known about the likely shading conditions on a system a reduced neighbourhood size with an appropriately designed cooling function could speed up the process of searching for the GMPP. In this study the geometric cooling function from (6) is used in this implementation.

The results show that the method converges to the 5%, 5V threshold in over 93% of cases considered regardless of the neighbourhood size. Figure 11 shows the average number of cases that converge based on the number of modules defining the neighbourhood for each system size. There is no clear trend here suggesting an optimal neighbourhood size to locate the region of the GMPP. All results show good convergence to the region of the GMPP regardless of the size of the neighbourhood. This may be due to the fact that even if a small neighbourhood is used, it dynamically moves within this implementation to be centered on the best known operating point enabling a wider range to be searched than if static boundaries were utilised. The results do show however that as the system becomes larger in terms of series-connected modules the average number of cases that converge reduces regardless of the initial searching neighbourhood size. As is supported by other results in this chapter, the larger system sizes may require additional steps in the search to reach higher convergence performance. When considered on a system size basis, and manufacturer basis, there is no clear trend suggesting which neighbourhood size should be selected. One key point of interest from these results however is that a satisfactory result can be obtained when using a small neighbourhood for searching relative to the operating voltage range of the system.

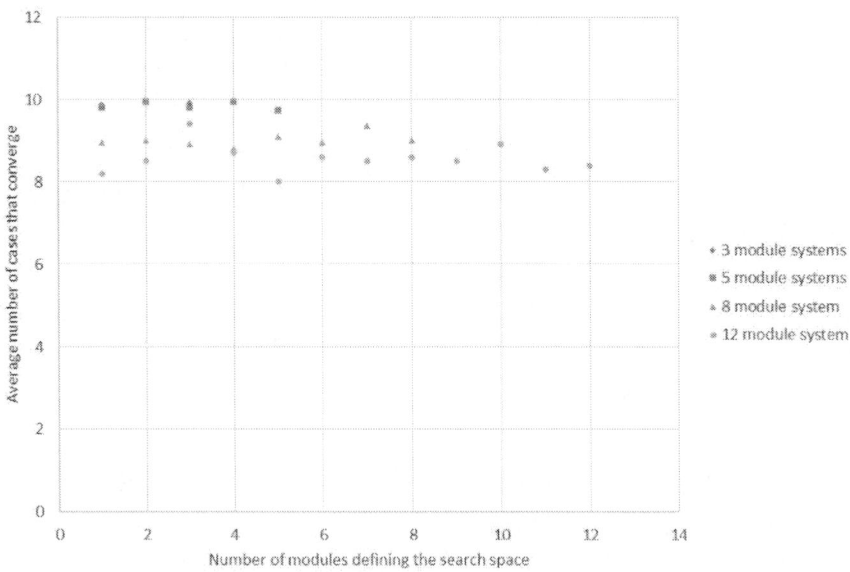

Figure 11. P-V characteristics for conditions leading to similar MPPs.

DISCUSSION

In this chapter the implementation of SA for tracking the MPP of PV systems has been explored. Parameters shown previously to have significance in the performance of the method have been explored on different system configurations based on differing numbers of series-connected modules and different manufacturer datasheet parameters. The performance of the algorithm is shown to be satisfactory relatively independently of the choice of parameters. However, the number of samples required in some of the implementations would be a barrier to implementation due to the rapidly changing environmental conditions a PV system may be exposed to.

As with other MPPT methods, there is a trade-off with the SA based MPPT method between the number of samples required and the tracking accuracy. Despite this, the method has good performance regardless of the parameters chosen.

The Lundy cooling schedule was shown to have the better performance across all systems and all parameters values when compared with the SQ implementation. Although the performance did reduce as the number of modules in series increased. Additionally the Lundy implementation required between 100 and 497 samples which is a significant barrier to implementation. The Lundy implementation may be most suitable for areas where shading is likely to change very slowly such as small systems where shading arises from an object in the environment.

The SQ algorithm had worse performance but required considerably fewer samples than the Lundy and neighbourhood reduction function implementations. Again the performance worsened as the number of modules in series increased for each parameter value. This suggests that these constants may be optimized for smaller systems with different constants necessary to improve the performance of larger systems.

The SA based MPPT method had good performance in locating the region of the GMPP regardless of the size of searching neighbourhood considered. This suggests that neighbourhood restriction could be a useful tool to improve the algorithm performance.

In terms of dynamically reducing the neighbourhood size, good results were observed with an arbitrary reduction parameter, however results suggest that this parameter could be tuned to improve the performance for particular system sizes.

Conclusion

This chapter has explored parameters which may impact the performance of the SA based MPPT method considering PV system made up of different configurations with different manufacturer specifications. General trends suggest that the Lundy cooling schedule has the best performance, although requires more samples in the searching time. The SQ implementation converges in most cases, and has the advantage of requiring fewer samples in its implementation. The neighbourhood size is shown to be quite independent of the searching performance, suggesting that if information is known about the likely shading conditions on the system this could be restricted to enable finer searching within an optimal range. Future work will involve additional studies into the impact of these parameters and establishing a framework to recommend to PV system designers the parameters they should adopt and tune when implementing a SA based MPPT method.

References

Ao, S. I. & Gelman, L. (2012). *Electrical engineering and intelligent systems*. Dordrecht: Springer. Retrieved from http:// utas.eblib.com.au. ezproxy.utas.edu.au/patron/FullRecord.aspx?p=972044&echo=1&userid=Sj1kVGdiDn6M6%2fR7j0drvw%3d%3d&tstamp=1385440319&id=42DBFBE6450C530FA4B58ED026C0A6FA333C561C.

BP 380 80 *W photovoltaic module datasheet*. (n.d.). Retrieved from http:// www.solarcharge.com.au/docs/off-grid-solar-panel-BP-80-Watt.pdf.

Bureau of Meteorology. (2012). *One Minute Solar Data*. Retrieved from http://www.bom.gov.au/climate/data/oneminsolar/about-IDCJAC0022.shtml.

Černý, V. (1985). Thermodynamical approach to the traveling salesman problem: An efficient simulation algorithm. *Journal of Optimization Theory and Applications*, *45*(1), 41–51. http://doi.org/10.1007/ BF0094 0812.

Chen, Y. M., Lee, C. H. & Wu, H. C. (2005). Calculation of the optimum installation angle for fixed solar-cell panels based on the genetic algorithm and the simulated-annealing method. *IEEE Trans. Energy Convers.*, *20*(2), 467–473. http://doi.org/10.1109/TEC.2004.832093.

Chong, E. K. P. & Zak, S. H. (2011). *An Introduction to Optimization*. Hoboken: Wiley. Retrieved from http:// utas.eblib.com.au.ezproxy.utas.edu.au/patron/FullRecord.aspx?p=708280&echo=1&userid=Sj1kVGdiDn6M6%2fR7j0drvw%3d%3d&tstamp=1385945814&id=0D8341254221CA24A0BCED34B1B9B7DEAB44FC49.

Coddington, P. (n.d.). *Simulated Quenching*. Retrieved from https:// cs.adelaide.edu.au/~paulc/teaching/montecarlo/node148.html.

Cohn, H. & Fielding, M. J. (1999). Simulated annealing: searching for an optimal temperature schedule. *SIAM Journal on Optimization*, *9*(3), 779–802.

Katsigiannis, Y. A., Georgilakis, P. S. & Karapidakis, E. S. (2012). Hybrid simulated annealing–tabu search method for optimal sizing of autonomous power systems with renewables. *IEEE Trans. Sustain. Energy*, *3*(3), 330–338. http://doi.org/10.1109/TSTE.2012.2184840.

Kobayashi, K., Takano, I. & Sawada, Y. (2003). A study on a two stage maximum power point tracking control of a photovoltaic system under partially shaded insolation conditions. In *IEEE Power Engineering Society General Meeting*, (Vol. *4*, p. 2617 Vol. *4*). Conference Proceedings. http://doi.org/10.1109/pes.2003.1271058.

Lyden, S. (2015). *A Simulated Annealing Global Maximum Power Point Tracking Method for Photovoltaic Systems Experiencing Non-Uniform Environmental Conditions*. University of Tasmania.

Lyden, S. & Haque, E. (2015a). A Simulated Annealing Global Maximum Power Point Tracking Approach for PV Modules under Partial Shading Conditions. *IEEE Transactions on Power Electronics*, *31*(6), 4171–4181. http://doi.org/10.1109/TPEL.2015.2468592.

Lyden, S. & Haque, M. E. (2014). Comparison of the Perturb and Observe and simulated annealing approaches for maximum power point tracking in a photovoltaic system under partial shading conditions. In *2014 IEEE Energy Conversion Congress and Exposition (ECCE)*, (pp. 2517–2523). IEEE. http://doi.org/10.1109/ECCE.2014.6953736.

Lyden, S. & Haque, M. E. (2015b). A hybrid simulated annealing and perturb and observe method for maximum power point tracking in PV systems under partial shading conditions. In *2015 Australasian Universities Power Engineering Conference: Challenges for Future Grids*, *AUPEC 2015*. http://doi.org/10.1109/AUPEC.2015.7324803.

Lyden, S. & Haque, M. E. (2016a). A comprehensive study of the key parameters of the Simulated Annealing method for maximum power point tracking in photovoltaic systems. In *2016 IEEE Power and Energy Society General Meeting (PESGM)*, (pp. 1–5). IEEE. http://doi.org/10.1109/PESGM.2016.7741788.

Lyden, S. & Haque, M. E. (2016b). A Simulated Annealing Global Maximum Power Point Tracking Approach for PV Modules under Partial Shading Conditions. *IEEE Transactions on Power Electronics*, *31*(6). http://doi.org/10.1109/TPEL.2015.2468592.

Lyden, S., Haque, M. E. & Xiao, D. (2014). Application of a Simulated Annealing technique for global maximum power point tracking of PV modules experiencing partial shading. In *40th Annual Conference of the IEEE Industrial Electronics Society (IECON)*, (pp. 1977–1983). http://doi.org/10.1109/IECON.2014.7048773.

Nourani, Y. & Andresen, B. (1998). A comparison of simulated annealing cooling strategies. *Journal of Physics A: Mathematical and General*, *31*(41), 8373–8385. http://doi.org/10.1088/0305-4470/31/41/011.

Patel, H. & Agarwal, V. (2008). Maximum power point tracking scheme for PV systems operating under partially shaded conditions. *IEEE Trans.*

Ind. Electron., *55*(4), 1689–1698. Journal Article. http://doi.org/10.1109/tie.2008.917118.

Poole, D. L. & Mackworth, A. K. (2010). *Artificial intelligence: foundations of computational agents*. Cambridge: Cambridge University Press. Retrieved from http://utas.eblib.com.au.ezproxy. utas.edu.au/patron/FullRecord.aspx?p=803005&echo=1&userid=Sj1kVGdiDn6M6%2fR7j0drvw%3d%3d&tstamp=1385685763&id=833606ADB7F8F5476BAA94498493653BC5D808DB.

Sera, D., Mathe, L., Kerekes, T., Spataru, S. & Teodorescu, R. (2013). On the perturb-and-observe and incremental conductance mppt methods for PV systems. *IEEE Journal of Photovoltaics*, *3*(3), 1070–1078. http://doi.org/10.1109/JPHOTOV.2013.2261118.

Soliman, S. A. H. & Mantawy, A. A. H. (2011). *Modern Optimization Techniques with Applications in Electric Power Systems*. Dordrecht: Springer. Retrieved from http:// utas.eblib.com.au.ezproxy.utas.edu.au/patron/FullRecord.aspx?p=884308&echo=1&userid=Sj1kVGdiDn6M6%2fR7j0drvw%3d%3d&tstamp=1385943415&id=91F46DF584B2658DF849263AF544E7A7B810750B.

Sutthibun, T. & Bhasaputra, P. (2010). Multi-objective optimal distributed generation placement using simulated annealing. In *International Conference on Electrical Engineering/Electronics Computer Telecommunications and Information Technology (ECTI-CON)*, (pp. 810–813).

World Energy Council. (2013). *World Energy Resources: Solar*. Retrieved from https://www.worldenergy.org/wp-content/uploads/2013/10/ WER_2013_8_Solar_revised.pdf.

Zäpfel, G., Braune, R. & Bögl, M. (2010). *Metaheuristic search concepts: a tutorial with applications to production and logistics*. Dordrecht: Springer. Retrieved from http://utas.eblib.com.au.ezproxy.utas.edu.au/patron/FullRecord.aspx?p=993620&echo=1&userid=Sj1kVGdiDn6M6%2fR7j0drvw%3d%3d&tstamp=1385690365&id=7DF90D70542BA0CB5FC6C5AE2D624BF5FDE1DE1E.

In: Simulated Annealing
Editors: Alex Scollen et al.
ISBN: 978-1-53613-674-6

Chapter 4

APPLICATION SIMULATED ANNEALING ALGORITHMS FOR OPTIMIZATION OF INFRARED HEATING STAGE IN THERMOFORMING PROCESS

***F. Erchiqui*[1,*] *and N. Nahas*[2]**
[1]Université du Québec en Abitibi-Témiscamingue
Rouyn-Noranda (Québec), Canada
[2]King Fahd University of Petroleum & Minerals,
Dhahran, Saudi Arabia

ABSTRACT

The quality of the molded product in thermoforming process depends on the distribution of intercepted infrared radiation by the thermoplastic sheet surface in oven, the type of polymeric sheet, the optical properties and the temperature of the radiant zones and the mold temperature. To ensure that the energy flux intercepted by the different zones of the polymer is minimal, we propose the application of simulated annealing

* Corresponding Author: fouad.erchiqui@uqat.ca.

algorithm to detect, from a fixed and random set of temperatures of the radiant zones of oven, the best temperatures that must be assigned to the heating zones. For numerical heating analysis, the nonlinear heat conduction problem is solved by a specific 3D volumetric enthalpy-based finite element method. An example of optimization of the heating stage of the high-density polyethylene (HDPE) grade sheet is presented.

Keywords: simulated annealing; genetic algorithm; optimisation; infrared heating; finite element; thermoforming

1. INTRODUCTION

The main processes for making plastic products are injection molding, extrusion, blow molding, thermoforming, and rotational molding. For its various applications, the processing conditions play an important role on the quality of a molded part. For thermoforming, the processing conditions are mainly related to the three steps. In first step, called Heating stage; allowing the transformation of the polymeric sheet from a glassy state to a rubbery state. For this stage, it is necessary that the distribution of the temperature in the polymeric material is included in the thermoforming temperature window. If we exceed this interval, the sheet, under the combined influence of temperature and gravity, can flow. Should the opposite occur, the material preserves its rigid state, and it's shaping becomes difficult to handle. The second step, called forming stage; allowing, with an appropriate force and a judicious choice of the mold, to transform the sheet, previously hot, into a finished product. For this stage, the plastic sheet stretches differentially against the mold surface with non-uniform distribution in thickness. The last step, called Cooling stage (solidification) where the polymeric material, in the form of the desired product, cools through its surface in contact with the mold. In this step, the difference in thickness (caused in the forming stage), cooling rate, and cooling time across the part surface may lead to different thermal stresses in the final part. These stresses, in their turns, can lead to part problems such crystallinity, warping, shrinkage, and part distortion. The prediction of thickness and residual stress distributions induced in the thermoformed part requires robust modeling of energy, momentum equations and fluid-structure interaction between the gas source and the viscoelastic plastic sheet [1-5]. However, the heating step has a first-rate impact on the quality of temperature distribution on plastic sheet. This temperature distribution depend depends on several variables such as properties of the sheet (thermo-physical and thermo-mechanical), heating time, optical properties of the radiant zones, etc. It is therefore critical to accurately predict sheet reheating in the thermoforming process [1-3].

From the point of view of numerical optimization, the methods are often based on continuous algorithms that use the techniques of differential calculus to locate optimum points. For example, the problem of trying to minimize thickness gradients in thermoformed parts via optimizing the heating has been discussed in the literature (different approaches have been adopted): McEvoy et al., [4] underscored the importance of optimizing the reheating stage in stretch blow molding. Monteix [7] optimized the processing parameters for the reheating phase of a 2-stage injection stretch blow molding process. Bordival et al. [8] proposed a simple analytical model, coupled with a nonlinear constrain optimization method (Sequential Quadratic Programming), for the optimization of infrared heating step for the thermoforming process. Boutaous et al., [9] developed a method to model and optimize radiant heating for plastic thermoforming using ray tracing. Nevertheless, the use of metaheuristic algorithms to optimize the thermoforming process has not yet been sufficiently clarified, particularly for the heating phase. Among the optimization methods that can be adapted to thermoforming, there is the meta-heuristic approach proposed by Erchiqui et al. [10]. In this approach, two iterative algorithms are proposed namely a simulated annealing algorithm and a genetic algorithm to produce a minimal difference between of the energies intercepted by the regions forming the thermoplastic structure. However, this method has not been coupled to the energy equation. It is for this purpose that the present work is oriented and the numerical impact of a meta-heuristic approach on the heating step in thermoforming.

From the point of view of numerical modeling, the classical methods of infrared heating of thermoplastics materials have been the subject of many studies in the literature. It is most often processed using the 1D, 2D and 3D finite element method [11-13], 1D and 2D finite difference method and 3D [14-15] and 3D control-volume method [16]. Also, when the plastic material undergoes a phase change, the numerical solution is more difficult due to the presence of one or more moving boundaries of the solid-liquid phases. In general, two approaches to this type of problem are used: i) solving the energy equations for the liquid and solid phases separately [17] and ii) solving simultaneously the energy equations for the liquid and solid phases (moving boundary problem: solid-liquid interface) by the enthalpy method [18].

The goals of the present study:

1) Determine the optimum temperatures to be assigned to a set of radiant zones from a set of randomly generated temperatures. For this, we consider simulated annealing (SA) [19] to reduce the differences in energy intercepted by different areas of a thermoplastic sheet. The SA is a probabilistic approach which is analogous to the metal cooling and freezing until a perfect crystalline structure. The algorithm simulates the metal cooling process by lowering the temperature of the system until it converges to a state with a minimum possible energy. The analogy of SA with a mathematical minimization was proposed by Kirckpatrick et al., [27]. The physical material states correspond to feasible solutions, the energy of a state to cost of a solution, the change of state to neighbouring solutions, the frozen state to heuristic solution and the temperature to a control parameter.
2) Study the impact of the temperatures of radiant zones obtained in i) on the temperature evolution of the thermoplastic sheet. To this end and for a more realistic modeling, the thermos-physical properties of plastic sheet are a function of temperature.

In addition, the thermoplastic material is considered opaque and the sagging deformation of the sheet, caused by combined effect of gravitational force and temperature, is not considered in this study.

2. Description Problem and Objective Function

The predicting the thickness distribution in the thermoformed part requires robust modelling of the energy and momentum equations. In particular, good modeling and optimization of the heating stage allows a better mass distribution in the thermoformed part, which reduces the number of rejections and improves productivity. For this, we consider in this introductory paper the use of simulated annealing (SA) to determine the optimum temperatures to be assigned to a set of radiant zones from a set of randomly generated temperatures. Then, with the use of the optimal temperatures obtained, we simulate the evolution of temperature within the plastic sheet. For use the SA approach, consider an infrared emitter, formed of a finite number of planar heating zones, and a flat surface S-shaped receiver (plastic-sheet). The intercepted radiation Q_i^j on cell i of the thermoplastic sheet surface S_i is calculated according to the geometric orientation S_j and temperature τ_j of oven heater cell j.

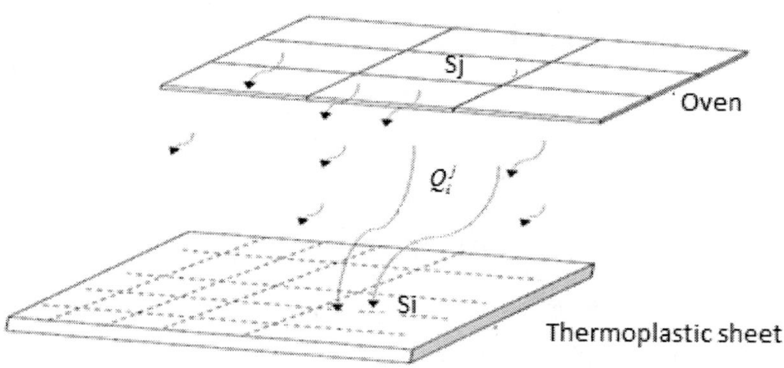

Figure 1. Heating stage: thermoforming process.

Given are m pre-specified available heating cells on an oven. For each cell j ($j = 1,...,m$), a temperature from set $T = \{T_1,...,T_m\}$ is assigned to minimize the standard deviation of the set of fractions of the energy received by the n thermoplastic sheet cells. To be more explicit, we proceed as follows:

Let x_{ij} be a decision variable such that $x_{ij} = 1$ if temperature $T_i \in T$ is assigned to cell j, and 0 otherwise. The expression of Q_{ij}^k is given by formula (1), for a given temperature T_k in $T = \{T_1,...,T_m\}$:

$$Q_{ij}^k = F_{ij} \frac{S_i}{S_j} \bar{\varepsilon}(T_k) \sigma T_k^4 \tag{1}$$

Let x_{ij} be a decision variable such that $x_{ij} = 1$ if temperature $T_i \in T$ is assigned to cell j, and 0 otherwise. It then follows that the total energy received by cell i of the thermoplastic sheet is:

$$E_i = \sum_{j=1}^{m} \sum_{k=1}^{q} Q_{ij}^k \times x_{kj}. \tag{2}$$

If we denote by $\bar{q}$ the average of the energy received by the n thermoplastic sheet cells, which is:

$$\bar{q} = \frac{1}{n} \sum_{i=1}^{n} E_i, \tag{3}$$

Then, the goal is to minimize the standard deviation of the energy received by the n cells of the thermoplastic sheet, which can be achieved by the following objective function:

$$\text{Minimize} \sqrt{\frac{1}{n} \sum_{i=1}^{n} \left(E_i - \bar{q}\right)^2}. \tag{4}$$

We note that the above optimization problem is an extended version of the quadratic assignment problem. Indeed, the problem we are considering here may be viewed as seeking a mapping between the set of the temperatures and the set of the oven cells such that objective function (4) is minimized. Therefore, the metaheuristic approach is well justified as the solving method of the optimization.

3. Simulated Annealing for Heating Stage in Thermoforming

Simulated annealing is an optimization method used for solving combinatorial minimization problems. A new solution is generated and compared against the current solution. The new solution is accepted as the current solution if the difference in quality does not exceed a dynamically selected threshold. The high-level overview of SA algorithm is given by the following steps:

1) Generate randomly an initial solution S_o and set $S \leftarrow S_o$.
2) Set initial temperature $T = T_{init}$ and cooling rate c ($0 \leq c \leq 1$).
3) Calculate the cost $f(S_o)$ of the initial solution S_o
4) Select randomly a new solution S' from the neighborhood space.
5) Calculate the cost $f(S')$
6) If $(f(S_o)- f(S')) > 0$ (for a minimization problem) then accept the new solution (i.e., $S \leftarrow S'$)
7) If $(f(S_o)- f(S')) \leq 0$ then accept the new solution with the probability $\exp(f(S_o)- f(S')/T)$.
8) Update the value of the temperature T
9) Repeat steps 4-8 until an acceptable solution is found or the maximum number of iterations is reached

The simulated annealing algorithm proposed to optimize temperature assignment is given in the reference [10].

4. Heating Stage in Thermoforming: Theory and Calculation

For application of the simulated annealing (SA) to determine the optimum temperatures to be assigned to a set of radiant zones, from a set of randomly generated temperatures, it is necessary to determine the radiation intercepted by the thermoplastic sheet when exposed to infrared radiation. This energy depends on the geometric orientation of the sheet and the arrangement of the zone heaters and their temperatures and optic properties. For this, in this section, we present the main formulas used to obtain the heat radiation intercepted by polymeric sheet. Then, for the study the impact of the temperatures of radiant zones on the temperature evolution of the thermoplastic sheet, we consider a specific 3D volumetric enthalpy-based finite element method.

4.1. Heat Radiation Intercepted by Polymeric Sheet

Let us suppose that the surfaces $\mathbf{S_i}$, of the heater, and $\mathbf{S_j}$, of the plastic sheet, are separated by a transparent medium. The amount of spectral radiation $Q_{ij,\lambda}$ leaving a finite area $\mathbf{S_i}$ that arrives at a finite area of $\mathbf{S_j}$ is given by the following equation [3]:

$$Q_{ij,\lambda}(\lambda, T_s) = F_{ij} \frac{S_i}{S_j} \varepsilon_\lambda(T_s) E_{\lambda,b}(\lambda, T_s) \quad (5)$$

F_{ij}, $E_{\lambda,b}(\lambda, T_s) = \sigma\, T_s^4$, $\varepsilon_\lambda(T_s)$, and T_s are, respectively, the view factor, the monochromatic emissive power of black-body source, the monochromatic emissivity of the plastic sheet and source temperature. σ is the Stefan-Boltzmann constant $\left(5.67 \times 10^{-8}\, W/m^2 K\right)$. The total of the radiation intercepted by the plastic surface may then be obtained by integrating over all wavelengths λ. For a Lambertian heater source (isotropic source) and considering an average source emissivity, $\bar{\varepsilon}$, of material, the total radiation intercepted by the plastic surface, eq. (5), becomes:

$$Q_{ij}(T_s) = F_{ij}\frac{S_i}{S_j}\bar{\varepsilon}(T_s)\sigma T_s^4 \tag{6}$$

Formula 6 requires the calculation the view factor is essential. The exact computation of this factor is very difficult. However, there are numerical and probabilistic techniques for its calculation. In this work, we have used a semi-analytical method proposed in [24].

4.2. Enthalpy Model for Energy Equation

Heat conduction and phase changes in the media are naturally described by the energy conservation law in terms of the volumetric enthalpy H(T) [22]:

$$\frac{\partial H(T)}{\partial T} = \nabla^2\theta(T) + Q_{wave} \tag{7}$$

where θ(T) and Q_{rad} are respectively the thermal conductivity integral and the internal volumetric heat generation of infrared radiation. The thermal conductivity integral, θ(T), and the volumetric enthalpy, H(T), are then defined in [22].

The advantage of using an enthalpy rather than a temperature-based formulation is that it simultaneously eliminates the doubling of the energy equation and the Stefan conditions [22].

To solve the problem, we introduce into eq. (7) the boundary condition, as follows:

$$\nabla\theta + h(T - T_\infty) - \boldsymbol{q}\cdot\boldsymbol{n} = 0 \tag{8}$$

where q [W/m^2] is the radiative heat flux incident, **n** is the outward normal to the surface, h [W/m^2/°C] is the surface heat transfer coefficient, and T_∞ is the temperature of the surrounding medium (air). The term h $(T-T_\infty)$ represents the convection heat transfer from the material to the environment. The incident heat flux depends on the source configuration and the position of the material.

4.2.1. Finite Element Formulation

Galerkin's finite element method is widely used for modeling heat transfer. Regardless of the spatial dimensionality of the heat problem, the finite element method leads to first-order equations of a rank equal to the total number of nodes on the geometric mesh. Assuming that the enthalpy H is defined uniquely in terms of the temperature T, we can express the Kirchhoff transform θ as a function of H. We then derive the following algebraic system:

$$\boldsymbol{C}\cdot\frac{\partial \boldsymbol{H}}{\partial \boldsymbol{t}}+\boldsymbol{K}\cdot\boldsymbol{H}+\boldsymbol{G}\cdot\boldsymbol{H}^2=\boldsymbol{R}+\boldsymbol{Q} \tag{9}$$

where C is the global matrix of heat capacity, K and G are matrices of thermal conductance, and (R+Q) is the nodal thermal load vector. To improve the numerical stability of (9), the capacitance matrix C is diagonalized (lumped). Although lumping introduces numerical diffusion, it eliminates undesirable oscillations generated by consistent mass matrices.

4.2.2. Implicit Time Integration Scheme

Numerical time approximation schemes are used mainly to obtain the transient response. These numerical integration schemes derive recursion relations that relate H(t) at a moment of time t to H(t+Δt) at another moment of time $t+\Delta t$. The solution is then solved step by step starting from the initial conditions at time $t=0$ until the desired duration of the transient response is calculated. The most common numerical schemes for the solution of eq. (9)

belong to the weighted Euler difference family of time approximations, as follows:

$$H^{n+\theta} = (1-\theta)H^{n} + \theta H^{n+1}, with \quad \theta = \frac{t-t_n}{\Delta t} \tag{10}$$

The parameter θ varies in the range [0–1]. The θ schemes are unconditionally stable when θ≤1/2 and O(Δt) are accurate, with the exception of the $O(\Delta t^2)$–convergent Crank–Nicolson scheme (θ=1/2). Setting θ=1 leads to the backward Euler (fully implicit) scheme, which is only first-order accurate, but very stable and hence ideally suited for integration. In the present study, we consistently use the implicit Crank–Nicolson scheme [23]. In this case, eq. (9) becomes:

$$\left(\boldsymbol{K}^{*}_{n+1} + \boldsymbol{G}^{*}_{n+1}\right)\cdot \boldsymbol{H}_{n+1} = \boldsymbol{K}^{*}_{n}\cdot \boldsymbol{H}_{n} + \boldsymbol{G}^{*}_{n}\cdot \boldsymbol{H}^{2}_{n} + \boldsymbol{R}^{*}_{n,n+1} \tag{11}$$

where K^*, G^*, and R^* are modified global matrices and H_{n+1} is the vector of global nodal enthalpies at moment t_{n+1}.

5. Application of Metaheuristics in Thermoforming

5.1. Application of MA for the Location of Temperature-Zone Heaters

In this section, the simulated annealing algorithm is used. For this, we have considered 36 (6x6) location of zone heaters. The initial temperatures of zone heaters are randomly selected. The search space exceeds 3.71×10^{41}. Tables 1 illustrate example of initial temperature distributions for radiant zones.

Table 1. Example of arbitrary temperature data for oven zones

i	*j*	*1*	*2*	*3*	*4*	*5*	*6*
1		420	400	410	400	490	490
2		520	505	520	510	590	560
3		600	600	600	610	630	640
4		700	700	700	780	765	710
5		830	800	800	850	830	830
6		900	910	920	950	950	990

For optimization, we applied the SA and ran simulations to set the parameters. Table 2 present the best results obtained respectively for example.

For illustration of convergence of algorithm, we have presented for the example, the value of the objective function versus the number of evaluated solutions in Figure 2. From this figure, we can see that the proposed algorithm converges fast. In fact, not more than 40,000 iterations are necessary to obtain a near optimal solution.

Table 2. Best solutions of example

i	*j*	*1*	*2*	*3*	*4*	*5*	*6*
1		990	600	830	850	630	920
2		505	600	520	520	420	800
3		830	640	610	800	590	700
4		830	400	780	600	400	910
5		710	490	410	700	510	490
6		950	560	900	700	765	950

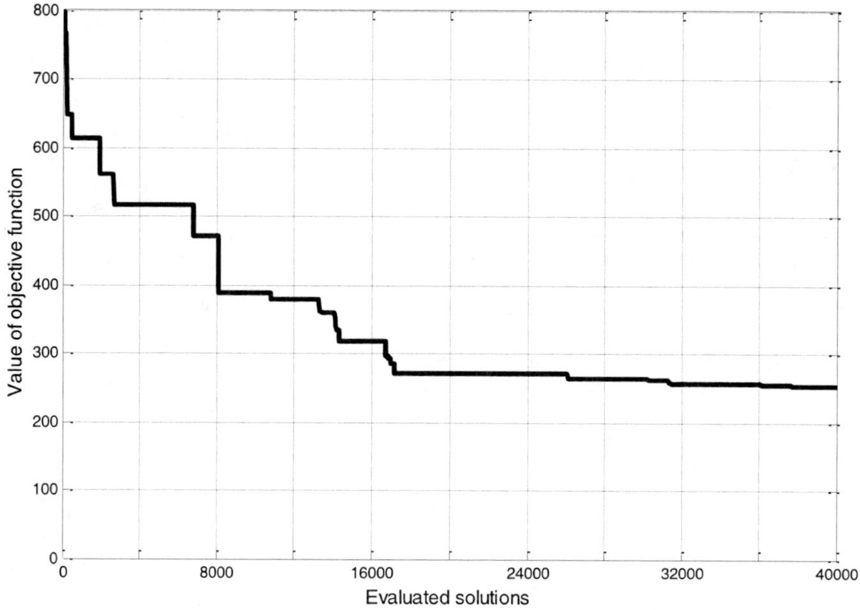

Figure 2. Convergence for the SA algorithm.

5.2. Application of SA for Heating Stage in Thermoforming

In this section, we present some simulation results associated with this heating stage of the thermoforming process without and with optimisation. For this, the dynamic finite-element method outlined in the previous section was implemented in the general-purpose finite-element code 'ThermoForm' developed by the principal author. All computations were performed in a single precision on a PC.

As an application, a 3D polymeric sheet material, polyethylene high density (HDPE), was considered. The sheet plastic was assumed to be homogeneous and isotropic and its thermal properties were functions of temperature (the density (ρ), specific heat capacity (Cp) and the conductivity (k) are calculated by the formulas provided in [10].

Table 3. Operating conditions for heating stage of thermoforming

Sheet dimensions	500 x 500 mm^2
Sheet thickness	3.0 mm
Upper oven dimensions	500 x 500 mm^2
Lower oven dimensions	500 x 500 mm^2
Initial distance between sheet upper oven	500 mm
Initial distance between sheet lower oven	500 mm
Number of heat zones (upper oven)	6x6
Number of heat zones (lower oven)	6x6
Heat zones temperatures (upper oven)	See table 1 to table 4
Heat zones temperatures (lower oven)	See table 1 to table 4
Initial sheet temperature	20 ^{0}C
Sheet heating time	20 sec
Temperature of ambient air T_{air}	50 ^{0}C
Effective emissivity of sheet ε_{eff}	0.7
Sheet heat transfer coefficient, front surface, $h_{air,}$(Oven)	5.0 W/m^2 /^{o}C
Sheet heat transfer coefficient, back surface, $h_{air,}$(Oven)	5.0 W/m^2 /^{o}C

The operating conditions for heating stage are given in Table 3.

For numerical modeling, the structure is meshed with identical hexahedra comprising eight nodes (1539 elements and 2000 nodes). Figure 3 illustrates the profile of view factor obtained by numerical modelling (Thermoform). This profile is symmetrical (since the sheet is placed between the two oven plates symmetrically with distance 500 mm to the upper and the lower oven plate).

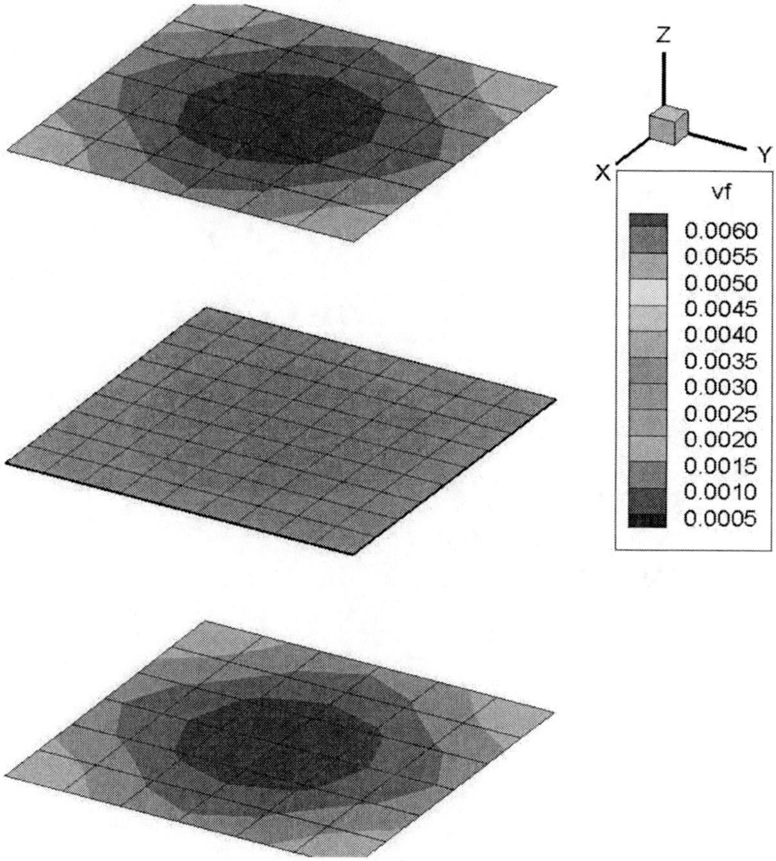

Figure 3. Distribution of view factor (plastic sheet and oven zones).

5.2.1. Case without Optimization

In Figure 4, we presented the distribution of radiation intercepted by the plastic surface for example. The temperature distribution induced by this infrared radiation, at the end of heating stage for front face, is presented at Figure 5. We note that the distribution of radiation energy and temperature are asymmetric. Effectively, since the temperatures of the various zones of the furnace are not identical and distributed in a random manner, this leads to asymmetric radiation heat flux on the plastic sheet. Consequently, the temperature distribution on and inside the sheet, with regard to the normal coordinate z, is also asymmetric.

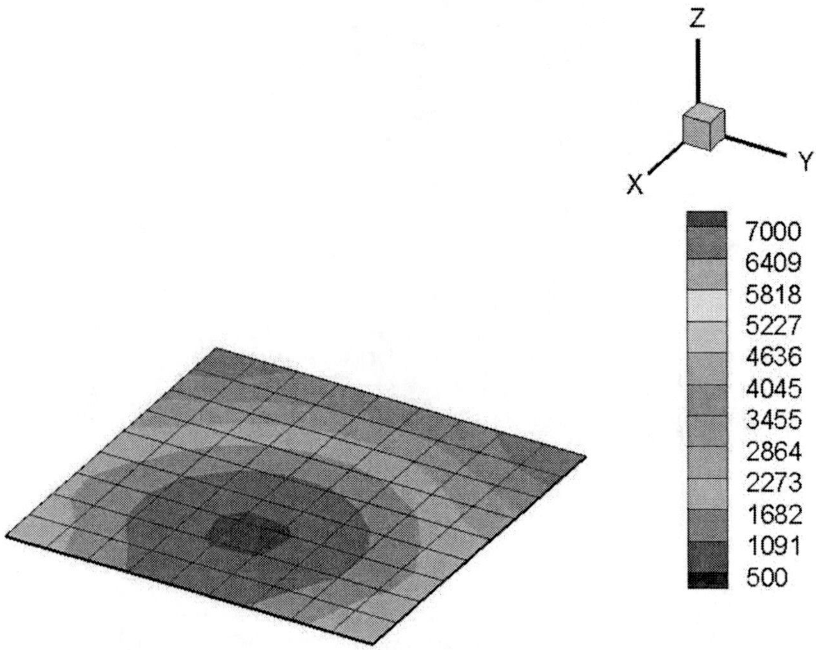

Figure 4. Distribution of radiation intercepted by the plastic surface for example without optimization.

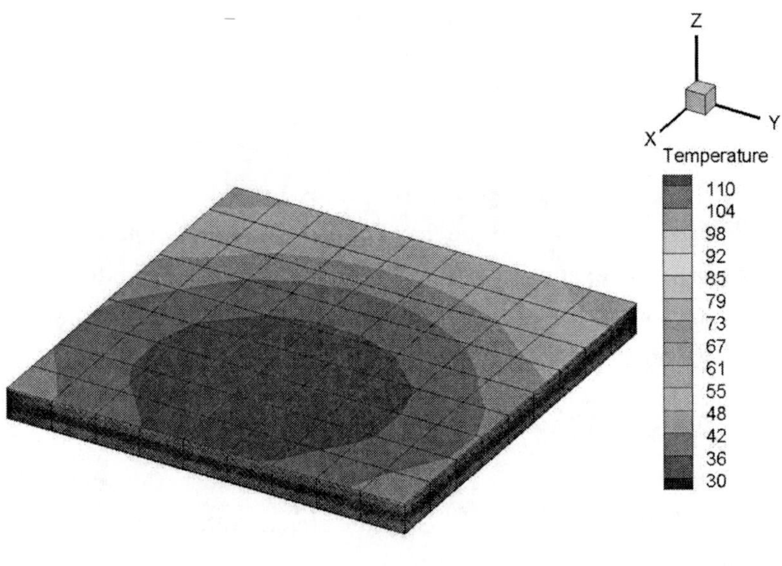

Figure 5. Distribution of temperatures for examples 1 and 2 without optimization.

5.2.2. Case with Optimization

In Figure 6, we presented the distribution of radiation intercepted by the plastic surface for example. The temperature distribution induced by this infrared radiation, at the end of the front face heating stage, is presented at Figure 7. We note that the distributions of radiation energy and temperature are almost symmetric (these distributions are better distributed than in the case without optimization). Effectively, even if the temperatures of the different zones of the furnace are not identical, their location, using the meta-heuristic optimization, is more organized. Consequently, the distribution of radiation energy intercepted by the plastic sheet is better distributed than in the case without optimization. Similarly, the temperature distribution on and inside the sheet, with regard to the normal coordinate z, is improved (almost symmetric).

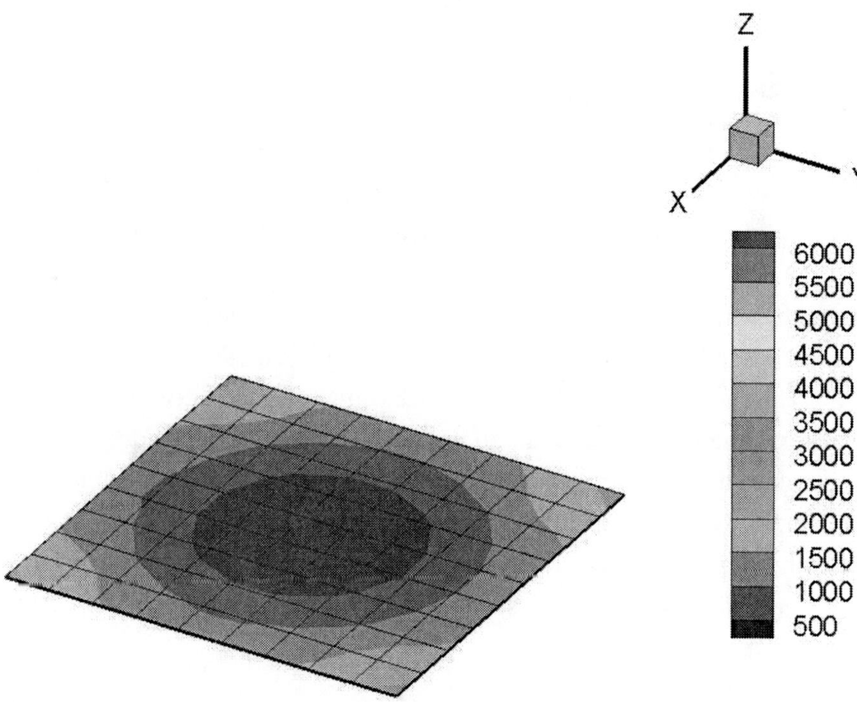

Figure 6. Distribution of radiation intercepted by the plastic surface for example with optimization.

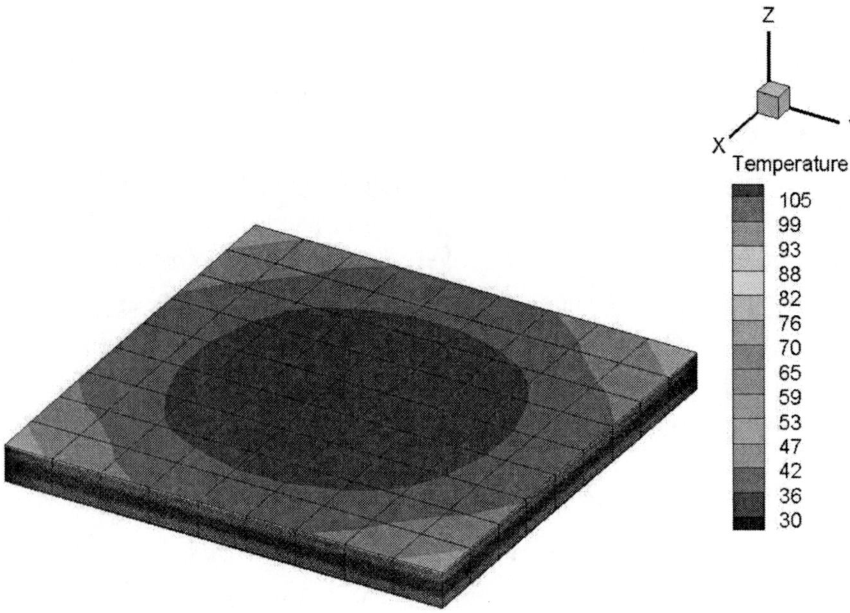

Figure 7. Distribution of temperatures for example with optimization.

For example (without and with optimization), we presented, in Figures 8-9, at the end of the heating stage process, the temperature profiles on the half-planes of symmetry XZ and YZ (middle lines of the plastic sheet) and on the thickness of the sheet (z direction) in Figure10. We observe that the areas exposed to weak radiative flux are characterized by weak temperatures, particularly the right zones of the plastic sheet. However, the results obtained with meta-heuristic optimization improve the quality of the temperature distribution in the plastic sheet. Note that these distributions of temperature have an impact on the strains and the stresses that can develop in the thermoplastic sheet. This phase of thermoforming is thus crucial for the forming and cooling phases. The temperature of the back face is hotter than that on the front and center faces.

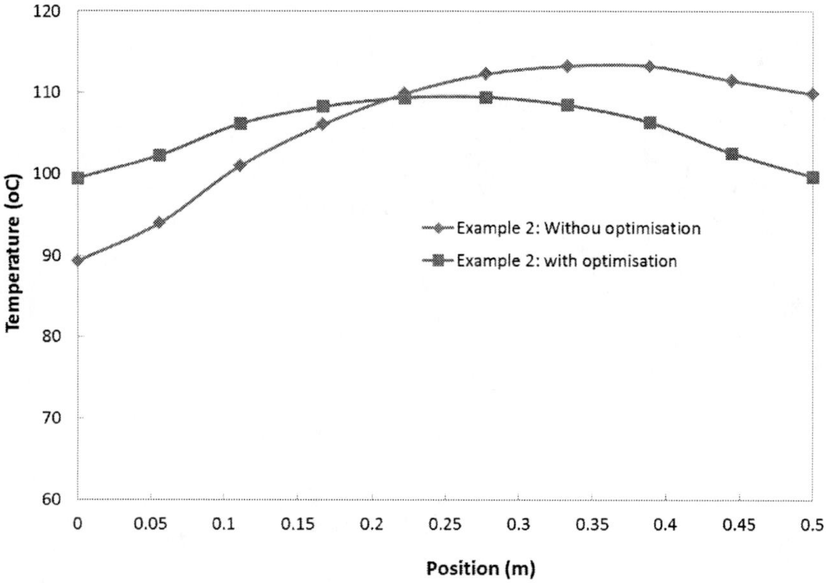

Figure 8. The temperature distribution on the half-planes of symmetry XZ (x direction).

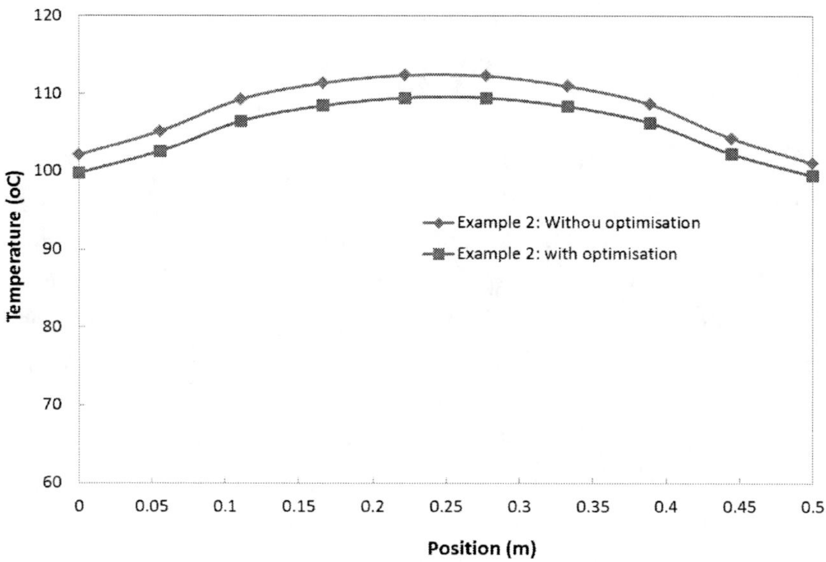

Figure 9. The temperature distribution on the half planes of symmetry YZ (y direction).

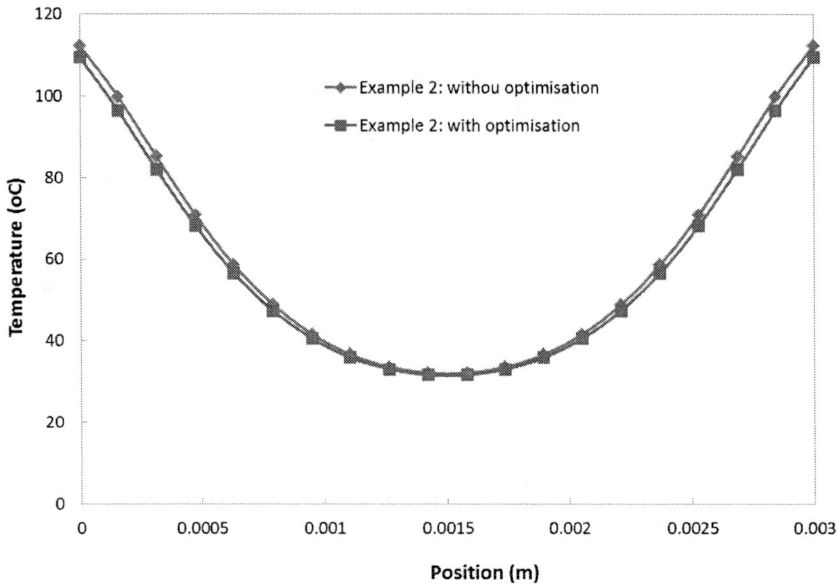

Figure 10. The temperature distribution on the half-planes of symmetry XY (z direction).

Conclusion

In this article, an investigation is realized out on the application of Simulated Annealing algorithms on the infrared heating step in thermoforming. For this, we initially created a random temperature distribution, representing infrared heater zones. Then, using an objective function, zones temperature assignments were optimised by minimising the variations in energy received between the various areas of the thermoplastic sheet.

An example of optimization of the heating stage of the high-density polyethylene (HDPE) grade sheet is presented. Optimization results obtained (solution quality and convergence) have shown that standard deviations of SA result are better for the example.

Also, the results obtained were used to determine the temperature evolution in the material. For this, the nonlinear heat conduction problem is solved by a specific 3D volumetric enthalpy-based finite element method.

We must underscore that the thermoplastic material is considered opaque and the sagging deformation of the sheet, caused by combined effect of gravitational force and temperature, was not considered in the optimization. This aspect is the subject of ongoing research.

In conclusion, by considering the small number of heating elements used in the thermoforming industry, the metaheuristic algorithms appear to provide effective optimisation of the infrared heating stage in thermoforming.

REFERENCES

[1] F. M. Duarte, J. A. Covas, IR sheet heating in roll fed thermoforming Part 1 –Solving direct and inverse heating problems. *Plastics, Rubber and Composites*, 31 (7) (2002) 307-317.

[2] F. Erchiqui F., A. S. Fotso-Talla, H. Kaddami, A numerical investigation of the use of novel melt processed PET-hemp fiber composites for thermoforming applications, *Polymer Engineering & Science*, 56 (9) (2016) 1021-1030.

[3] F. Erchiqui F., M. Souli, R. Ben Yedder. Nonisothermal finite-element analysis of thermoforming of polyethylene terephthalate sheet: Incomplete effect of the forming stage, *Polymer Engineering & Science*, 47 (12) (2007) 2129 -2144.

[4] J. P. McEvoy, C. G. Armstrong, R. J. Crawford. Simulation of the stretch blow molding process of PET bottles, *Advances in Polymer Technology*, 7 (4) (1998) 339-352.

[5] F. Erchiqui, M. Souli, T. Kanit, A. Imad, Characterization of Polymeric Membranes Under Large Deformations Using Fluid-Structure Coupling, *International Journal of Applied Mechanics*, 7 (5) (2015) 1-18.

[6] J. F. Agassant, J. M. Haudin, Book: *Mise en forme des polymères thermoplastiques. Matières plastiques, Propriétés, mise en forme et applications industrielles des matériaux polymères* [*Shaping thermoplastic polymers. Plastics, Properties, shaping and industrial applications of polymer material*], (2012) 53-110.

[7] S. Monteix, PhD dissertation, *École des Mines de Paris* [*School of Mines of Paris*] (France), 2001.

[8] M. Bordival, F. M. Schmidt, Y. Le Maoult, S. Monteix, Conference: *Proceedings of 8th ESAFORM Conference on material forming*, January 2005.

[9] M. Boutaous, P. Bourgin, D. Heng, D. Garcia, Optimization of radiant heating using the ray tracing method: Application to thermoforming, *Journal of advanced Sciences*, 17 (1/2) (2005) 135-139.

[10] F. Erchiqui, N. Nahas, M. Nourelfath, M. Souli, Metaheuristic algorithms for optimisation of infrared heating in thermoforming process, *International Journal of Metaheuristics*, 1 (3) (2011) 199-221.

[11] M. A. Thrasher, Finite Element Modeling of Radiant Heating on Polymeric Materials. *SPE ANTEC Tech, the Plastics Technology Conference*, (1996) 854-857.

[12] K. Hartwig, *Simulation of the stretch-Blow molding process and characterization of the relevant material behavior*, Dissertation, RWTH, Aachen (1996).

[13] F. Erchiqui, I. Hamani, A. Charrette, Modélisation par éléments finis du chauffage infrarouge des membranes thermoplastiques semi-transparentes, International Journal of Thermal Sciences, *International Journal of Thermal Sciences*, 49 (1) (2009) 73-84.

[14] G. Venkateswaran, M. R. Cameron, S. A. Jabarin. Effect of Temperature Profiles Trough Preform Thickness on the Properties of Reheat-Blown PET Containers. *Advances in Polymer Technology*, 17 (3) (1997) 237-249.

[15] F. M. Duarte, J. A. Covas, IR sheet heating in roll fed thermoforming Part 1 –Solving direct and inverse heating problems. *Plastics, Rubber and Composites*, 31 (7) (2002) 307-317.

[16] F. M. Schmidt, Y. Le Maoult, S. Monteix, Modelling of infrared heating of thermoplastic sheet used in thermoforming process, *Journal of Materials Processing Technology* 143–144 (2003) 225–231.

[17] H. Hu, S. A. Argyropoulos, Modelling of Stefan problems in complex configurations involving two different metals using the enthalpy method. *Modelling and Simulation in Materials Science and Engineering*, 3 (1) (1995) 53-64.

[18] B. Nedjar, An enthalpy-based finite element method for nonlinear heat problems involving phase change, *Computers & Structures*, 80 (1) (2002) 9–21.

[19] C. Koulamas, S. R. Antony, R. Jaen, A survey of simulated annealing applications to operations research problems, Omega *International Journal of Management Science* 22 (1) (1994) 41–56.

[20] J. H. Holland, *Control and Artificial Intelligence*, Ann Arbor, MI: University of Michigan Press, 1975.

[21] Burkard, R., Dell'Amico M., Martello S.: Assignment Problems, revised reprint, *SIAM Book*, ISBN 978-1-611972-22-1, 2012.

[22] F. Erchiqui, Z. Annasabi Z, M. Souli, F. Slaoui-Hasnaoui (2015), 3D numerical analysis of the thermal effect and dielectric anisotropy on thawing frozen wood using microwave energy, *International Journal of Thermal Sciences*, 89 (2015) 58-78.

[23] M. K Dokainish., K. Subbraj, Asurvey of direct time-integration methods in computational structural dynamics, *Computer & Structures*, 32(6) (1989) 1371–1386.

[24] F. Erchiqui, N. G. Dituba, Analyse comparative des méthodes de calcul des facteurs de formes pour des surfaces à contours rectilignes, *International Journal of Thermal Sciences*, 46 (3) (2007) 284-293.

[25] R. Siegel, J. R. Howel, *Thermal Radiation Heat Transfer*. 3rd Ed Hemisphere / Taylors and Francis, Washington, 1992.

[26] D. Whitley, J. Kauth. *Technical Report* CS-88-101, Colorado State University, Fort Collins, CO, 1988.

[27] S. Kirkpatrick, C. D. Gelatt and M. P. Vecchi, Optimization by simulated annealing, *Science*, 220 (4598), 671-680, 1983.

In: Simulated Annealing
Editors: Alex Scollen et al.
ISBN: 978-1-53613-674-6

Chapter 5

SIMULATED ANNEALING TECHNIQUES APPLIED TO POROUS MEDIA RECONSTRUCTION

***Alysson Nunes Diogenes, PhD*[1*] *and Celso Peres Fernandes, PhD*[2]**
[1]Department of Mechanical Engineering
Universidade Positivo, Curitiba, PR, Brazil
[2]Department of Mechanical Engineering
Universidade Federal de Santa Catarina, Florianopolis, SC, Brazil

Abstract

This chapter presents three Simulated Annealing applications, as well the Simulated Annealing theory. To reconstruct a porous media from two-dimensional (2D) images to three-dimensional (3D) images, several parameters such as the spatial correlation must be preserved. Otherwise, the 2D grain size distribution is known to be different from the 3D grain size distribution. Therefore, it is exposed a Simulated Annealing method to correct the grain size distribution from 2D to 3D. Likewise, it is presented two porous media reconstruction methods using the Simulated Annealing optimization method in three-dimension (3D). Measuring geometrical and topological parameters in (2D), it is possible to generate a 3D microstructure model. The methods use different SA movements: one is based on pixels (for 2D) or voxels movements (for 3D) (PSA) and the other moves objects: circles (2D) or spheres (3D) (OSA). Both of them

*Corresponding Author: diogenes@up.edu.br

preserve the spatial correlation function and the PSA method also preserves the linear chord distribution and complementary distance distribution parameters.

PACS: 05.45-a, 52.35.Mw, 96.50.Fm

Keywords: simulated annealing, porous media reconstruction, object-based, pixel-based

1. Introduction

The characterization of porous materials constitutes a problem of great scientific and technological interest. The determination of macroscopic transport properties of these materials has application in several areas, such as soil, oil exploration and construction and civil engineering, for example.

The developments of the present work are applied for the characterization of porous media. The geometry of the porous microstructures can present a high degree of complexity, so there are several properties, geometric or physical, that characterize these materials. Among these properties, it can be mentioned: porosity, pore size distribution, intrinsic permeability, relative permeability curves and electrical parameters.

For most cases, it is assumed homogeneity and isotropy statistics so 3D models will be generated from 2D images, retaining various information such as porosity, spatial correlation and linear chordal distribution of the pore and/or solid phases.

The method based on the truncation of Gaussian fields [Adler et al., 1990,Liang, 1997] is currently the most used for the 3D reconstruction of porous microstructures. However, this method presents problems regarding the conservation of long distance connectivity for environments with strong spatial organization, for example, porous media with low porosity and high intrinsic permeability. In fact, this method preserves only second-order statistics (porosity and two-point correlation), which is insufficient for structures with strong spatial organization. Some methods have been proposed, such as the Superposed Spheres [dos Santos et al., 2002].

Several researchers have been studying porous media characterization and their physical and geometrical properties. [Berryman and Blair, 1986] studied the use of spatial two-point correlation to estimate a porous media permeability. [Narayanan and Deans, 1988] worked on a flow model based on conventional X-Ray images acquisition. [Torquato, 1992] proposed the Lineal-Path function, which [Yeong and Torquato, 1998] and [Derossi et al., 2014] applied to porous media reconstruction with stochastic methods. Other researchers as [Ioannidis and Chatzis, 2000], [Manwart et al., 2000], [Talukdar and Torsaeter, 2002], [Bekri et al., 2000] and [Bekri et al., 2003] continued the research in reconstructing porous media with Simulated Annealing techniques, investigating for new parameters to adopt as cost function. An innovation was proposed by [Politis et al., 2008], which studied a hybrid stochastical approach. Likewise, [Bostanabad et al., 2016] used stochastic methods to characterize the porous media, instead of using the method to reconstruct it.

On the other hand, others researchers such as [Biswal et al., 1998], [Venkatarangan, 2000] and [Pazdniakou and Adler, 2013] worked on the quality of the reconstructed porous media, comparing bidimensional (2D) parameters with the three-dimensional (3D) parameters measured in the reconstructed porous media.

As shown, Simulated Annealing methods have been used extensively to reconstruct, characterize and optimize porous media. This chapter presents three applications of Simulated Annealing techniques. The first is a technique to correct a grain size distribution to reduce the difference between the 2D and 3D grain size distributions. The methods use different simulated Annealing approaches to reconstruct porous media: one is based at pixels (for 2D) or voxels movements (for 3D) (PSA) and the other moves objects: circles (2D) or spheres (3D) (OSA). Both of them preserve the spatial correlation function and the PSA method also preserves the linear chord distribution and complementary distance distribution parameters.

2. Porous Media Geometrical Parameters to Use as Simulated Annealing Cost Functions

One way of characterizing microstructures is from digital images of flat sections of the material. From these images, colored or gray levels, segmentations are made, and two phases are defined, solid and porous. The segmentation process for porous media images consists of the separation of the pore and solid phases. In the present case, the pore phase is colored white and the solid phase colored black. In the color images, from the histogram of the image, a threshold was established for each color component. In the Figure 1, a colored image of a flat section of a porous media is shown. In Figure 2, the binary image corresponding to the previous image is displayed.

Figure 1. Image of a flat section of a porous media obtained by light microscopy.

After segmentation, several parameters can be measured through the image. These parameters are intended to describe the geometry and topology of the image so that it is possible to reconstruct the medium in three dimensions. Among them, it can be mentioned porosity, autocorrelation function, linear chord function, lineal path function and others.

Another way is to use X-Ray microtomographic images. This fully characterizes the porous media for a defined scale and generates a 3D image.

After the image is acquired and segmented, some geometrical parameters can be measured.

2.1. Autocorrelation Function

In a binary (segmented) image, a phase function $Z_p(\vec{x})$ related to the porous phase is defined as:

$$Z_p(\vec{x}) = \begin{cases} 0 & for\ \vec{x} \in M \\ 1 & for\ \vec{x} \in P \end{cases} \tag{1}$$

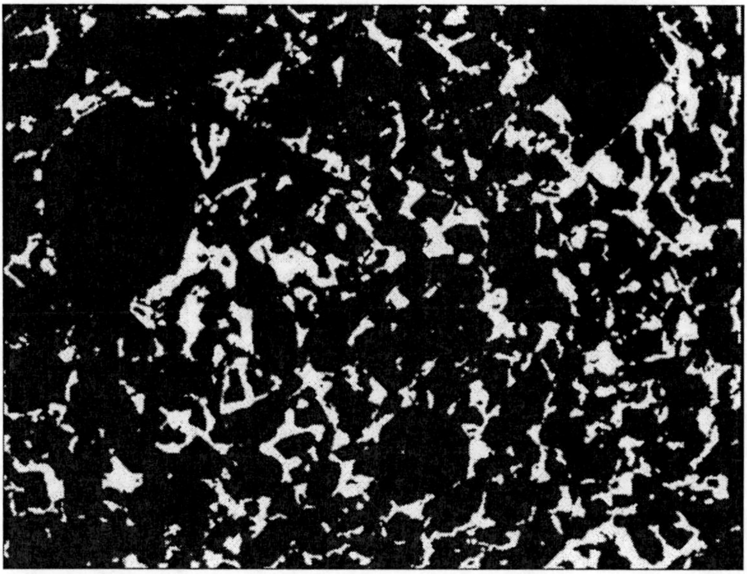

Figure 2. Binary image corresponding to the gray image of Figure 1 with blank pore phase.

where $\vec{x} = (i, j)$ denotes a position vector relative to an arbitrary origin. Once the phase function has been defined, and the hypothesis of a statistically homogeneous medium is adopted, the porosity, which is the volumetric fraction of the pore phase:

$$\phi = \langle Z_p(\vec{x}) \rangle \tag{2}$$

where $\langle\rangle$ represents the average for the image domain under consideration.

The correlation function for the porous phase is defined as:

$$C_p(\vec{u}) = \langle Z_p(\vec{x}) \cdot Z_p(\vec{x} + \vec{u}) \rangle \tag{3}$$

For each arbitrary displacement $\vec{u}$ in the image plan. Moreover, with the isotropic medium hypothesis, the correlation will only depend on $u = |\vec{u}|$, and:

$$C_p(\vec{u}) = \langle Z_p(i, j) \cdot Z_p(i + u, j) \rangle \tag{4}$$

The correlation function can be determined by moving the image over itself, in the vertical or horizontal direction in multiple pixels, by calculating the frequency of results associated with the intersection of two pixels belonging to the same phase.

Correlation provides the probability that two pixels separated by a distance u belong to the same phase. The statistical values of the correlation range from ϕ to $u = 0$, to ϕ^2 for large displacements. One can see an example of autocorrelation in Figure 3.

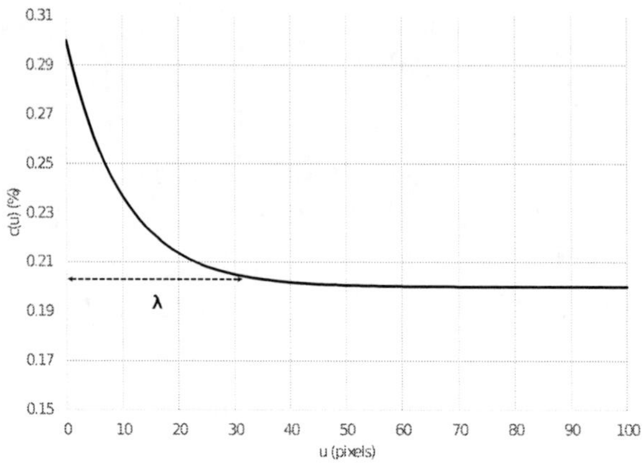

Figure 3. Autocorrelation function curve example.

2.2. Pore and Grain Size Distributions

The mathematical morphology allows one to extract morphological and topological information of the objects that compose an image, being based on an element of analysis called structuring element [Coster, 1989]. The structuring element is a discrete object, with defined shape and size, for comparison with the objects that make up the image. The result of the application of the operations of mathematical morphology depends, fundamentally, on the structuring element, that is, on the geometric form of the structuring element.

The basic operations of mathematical morphology are erosion and dilation [Serra, 1983]. The combination of these operations results in the opening operation, when erosion is followed by dilation, or in the closing operation when a dilation is followed by erosion. By using successive aperture operations with increasing structural member size, the cumulative distribution of pore sizes, or solids, is obtained in an image.

The study of the distribution of object sizes is commonly called granulometry. The term comes from the technique of determining the grain size distribution of materials such as sand using different mesh sieves.

The opening operation can be seen as the union of the balls, implanted in all the pixels of the objects, and which do not intercept with the complementary phase. The opening with a ball of size r presents analogy with the technique with sieve of physical mesh of radius r. In the case of the technique with sieves, the measuring entity is the mass of the objects, and, in the case of images submitted to the aperture, is the surface area or the volume of the objects. In the figure 4, an image is shown whose white phase is subjected to an opening operation with a ball of a given radius r, resulting in the image of Figure 5. The opening operation has the effect of "eliminating" objects smaller than the ball used, as well as "eliminating" the roughness of objects and nicks (throats). In analogy with the technique with physical sieves, it can be said that the "eliminated" objects passed through the mesh sieve r.

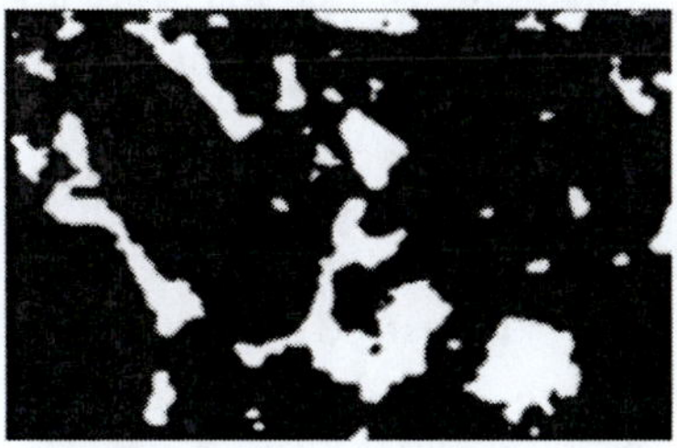

Figure 4. Porous media image.

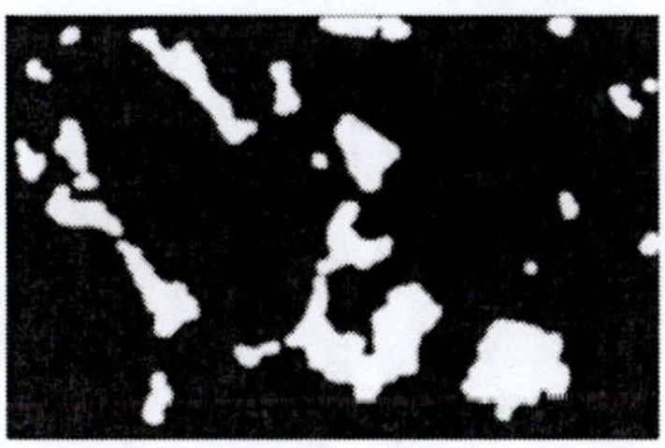

Figure 5. Porous media image after morphological opening.

The function of pore size distribution is determined by performing a sequence of aperture operations on a binary image with a structuring element of increasing size. For the determination of the pore distribution in 2D binary images, the unit of measurement consists of the pore surface area. Thus, the cumulative pore distribution will be given by:

$$F(D) = \frac{\phi - \phi(D)}{\phi} \tag{5}$$

where $F(D)$ is the fraction of pores with radius smaller than or equal to D, ϕ is the porosity and $\phi(D)$ is the porosity after the opening with a ball with diameter D.

It is also possible to observe the differential distribution of pores in the following:

$$F_d(D) = \frac{dF(D)}{dD} \tag{6}$$

Figure 6 shows the comparison between a cumulative pore size distribution and its differential equivalent. Notice the scale change.

In this chapter, balls were used in the $d_{3\text{-}4}$ metric system [Coster, 1989] to determine the pore size distribution. These balls were chosen because it has good results with a smaller computational cost if compared to balls that are similar to circular shapes. [Fernandes, 2002].

The same procedure can be used to obtain a 3D pore size distribution, changing only the metric system from $d_{3\text{-}4}$ to $d_{3\text{-}4\text{-}5}$ [Coster, 1989].

If the analyzed phase is the grain, instead of the pore, it can be said the grain size distribution. On the other hand, if instead of the grain size, the distance to the background measured using the [Coster, 1989] methodology, it is possible to measure a different parameter, called "Distance to the Background Image distribution" (DBI).

This parameter has much useful information, e.g., the curvature of the contact surface between the grains and pores . For this reason, it is an interesting candidate for cost function in SA algorithms.

2.3. Distance to Background Image Distribution (DBI)

Looking at the Figure 7b, it can be seen that the image has several distance to the background values. By measuring the frequency for each of the values, and normalizing by the higher frequency, an DBI distribution can be generated for

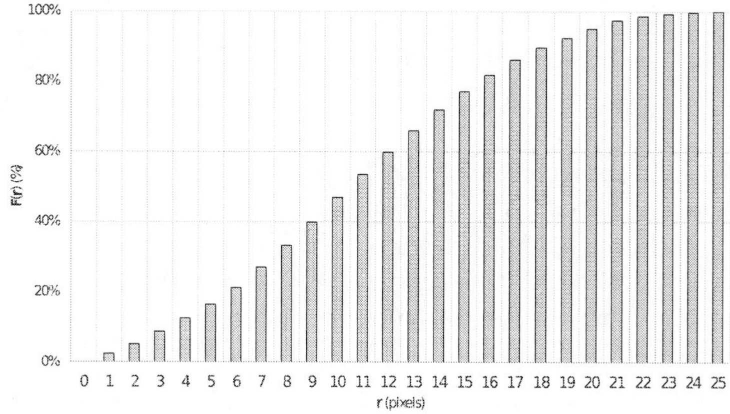

(a)

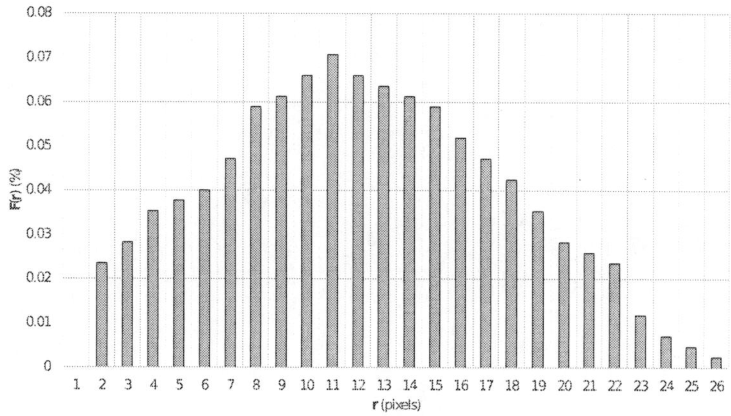

(b)

Figure 6. Comparison between the (a) accumulated and (b) differential pore size distribution.

a given metric. This parameter means that there are two images, such as equivalent pore size distribution objects. This parameter differs from the pore/grain size distribution as follows:

When calculating a pore size distribution by mathematical morphology, balls greater than or equal to $3r$ are calculated, due to morphological operations of erosion and dilation. For example, for the metric $d_{3\text{-}4}$, the radius 1 will comprise the DBI value 3, while the radius 2, the values 3, 4, and 6.

Thus, the pore size distribution does not differentiate the pixels of value 4 from those of value 6, for example. This effect increases as the radius of the ball increases.

Therefore, it is concluded that the DBI distribution can be considered a generalization of the pore size distribution for a given metric.

The sequential algorithm for obtaining DBI for an image $Im(i, j)$ of size MxN is described by [Chassery and Montanvert, 1991].

Figure 7 presents an example of a binary image and its corresponding distance-encoded image is exposed.

0	0	0	0	0	0	0	0
0	0	1	1	1	1	0	0
0	1	1	1	1	1	1	0
0	1	1	1	1	1	1	0
0	1	1	1	1	1	1	0
0	0	1	1	1	1	0	0
0	0	1	1	0	0	0	0
0	0	0	0	0	0	0	0

(a)

0	0	0	0	0	0	0	0
0	0	3	3	3	3	0	0
0	3	4	6	6	4	3	0
0	3	6	8	8	6	3	0
0	3	6	7	6	4	3	0
0	0	4	4	3	3	0	0
0	0	3	3	0	0	0	0
0	0	0	0	0	0	0	0

(b)

Figure 7. A binary image and its DBI coding with metric $d_{3\text{-}4}$: (a) original image, (b) image resulting from the transformation.

Figure 8 illustrates a comparison between these parameters for the metric $d_{3\text{-}4}$ for the image of the Figure 7b. It is observed that the differential pore size distribution varies with multiples of 3, its base value, while that of IDF varies according to all values of the transformation. This allows for greater refinement when characterizing the image.

This parameter aims to preserve the spherical or rhomboid shape of the grains, and in this chapter, its measurement must be performed on the grains of the image that one wishes to reconstruct.

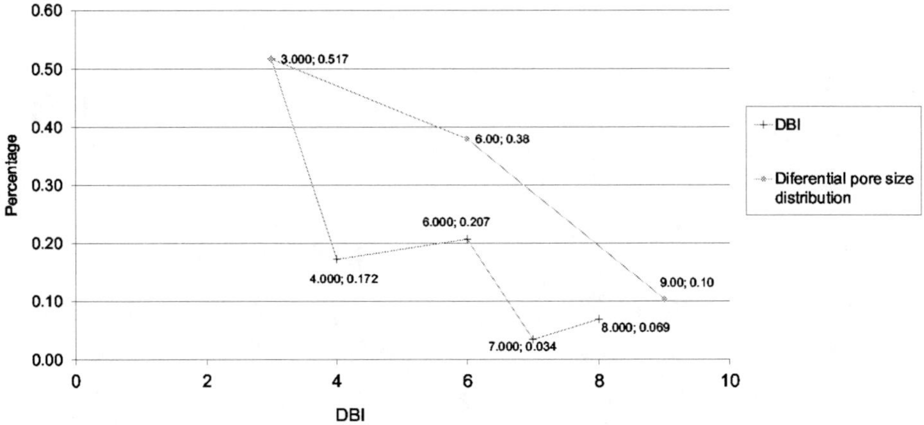

Figure 8. Comparison between the differential pore size distributions and the DBI distributions for the Figure 7b.

2.4. Chord Size Distribution or Linear Chord Distribution

For a 2D binary image, there is a chord distribution associated with each phase $p_i(z), (i = 1, 2)$. The quantity $p_i(z)dz$ is defined as the probability that a chord (a line segment with end at the phase interfaces) chosen at random in the phase i have length in range $[z, z + dz]$ [Torquato and Lu, 1993].

This parameter is calculated by randomly "tossing" a large number of chords of infinite size (larger than the image size) in the image in which this parameter is to be measured. Figure 9 illustrates this procedure. For each tossed chord, the amount of sections of chords are located entirely in phase 1 with length in the interval $[z, z + dz]$ are counted. If N is the total number of chord sections of phase 1, then:

$$p_1(z) = \frac{N_1(z)}{N} \tag{7}$$

The quantity $p_2(z)$ is defined in an analogous manner, and so on for the other chord sizes.

This parameter is used to estimate the connectivity of the phase in question.

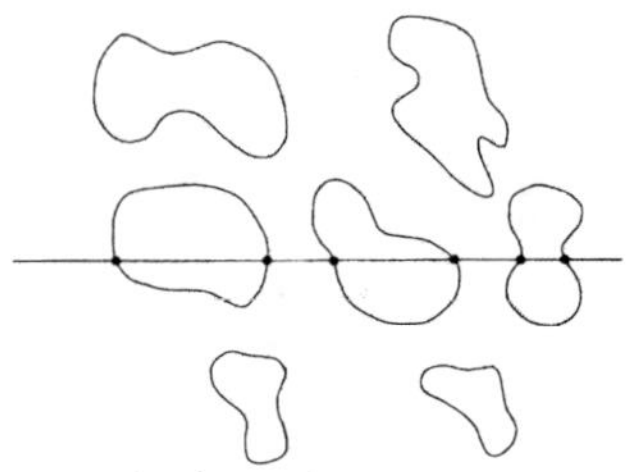

Figure 9. Illustration of the procedure for calculating chord size distribution.

3. Simulated Annealing Theory

This section describes the Simulated Annealing, as its mathematical model and the methods of reconstruction presented in this chapter.

3.1. Simulated Annealing (SA)

The SA algorithm derives from the observation that the solution of large-scale combinatorial problems is analogous to the baking of solids in the field of condensed matter physics [Metropolis et al., 1953, Kirkpatrick et al., 1983]. The purpose of this process is to reduce the temperature of a system to its minimum energy. Such energy can be seen as a cost function to be optimized. The process begins with a molten material that gradually has its temperature reduced, analogous to a process having its energy decreased. If its temperature is reduced too quickly, the process "freezes" in a sub-optimal state. A gradual reduction in temperature leads the system to better states in terms of energy decreases.

The initial temperature of the annealing process is the point at which all particles of the solid are randomly arranged. As the cooling starts, the solid must reach what is known as thermal equilibrium before cooling continues [Metropolis et al., 1953]. If the temperature is lowered before this, defects in the structure of the material will appear and the resulting crystal will not correspond to a minimum state of energy (i.e., frozen at a local minimum). A formal definition of the thermal equilibrium is given below:

Definition 1. *Thermal Equilibrium is achieved during the physical cooking process when the solid has the probability of being in a state of energy E according to the equation:*

$$P_r\{\mathbf{E} = E\} = \frac{1}{Z(T)} \cdot e^{\frac{-E}{k_B T}} \tag{8}$$

where $Z(T)$ is a temperature dependent normalization factor called the partition function and k_B is the Boltzmann constant [Metropolis et al., 1953]. The probability distribution exposed in Eq. 8 is also called the Boltzmann distribution. As it can be seen, only minimal energy states have a nonzero probability when the temperature tends to zero.

The simulation of Metropolis Monte Carlo [Metropolis et al., 1953] can be used to simulate the cooking process at a fixed temperature T. The Metropolis method randomly generates a sequence of states for the solid at a given temperature. The state of a solid is characterized by the positioning of its particles. A new state or perturbation state is generated by randomly moving a particle by a small distance. Acceptance or rejection of this new state is determined by Metropolis acceptance criteria, as described below:

If the disturbance state j has energy lower than the current state i, the Metropolis process continues from the perturbation state. If the energy of the disturbance state is greater than that of the current state, the probability of the continuation of the disturbance state is given by

$$P_r\{X_k = j|X_{k-1} = i\} = e^{\frac{-\Delta E_{ij}}{k_B T}} \tag{9}$$

where ΔE_{ij} is the energy difference between state j and state i. This rule is commonly known as the Metropolis criterion.

After a large number of states generated and subject to the Metropolis criterion, the temporal distribution of the states will approximate the distribution of Boltzmann [Metropolis et al., 1953].

3.1.1. Simulated Annealing Algorithm

[Kirkpatrick et al., 1983, Cerny, 1985] independently proposed that it is possible to simulate the annealing process by generalizing the Metropolis simulation to generate a sequence of solutions with the purpose of solving combinatorial optimization problems. The solutions to these problems are analogous to that of the solid state during the annealing process with the corresponding cost of the solution as being analogous to the solid state energy. Since temperature does not have a conceptual equivalent in most combinatorial systems to be optimized, it is modeled as a control parameter for the use of the SA algorithm. This algorithm can be described as a sequence of Metropolis simulations performed at decreasing temperature values.

In this traditional form, SA starts with a high starting temperature T_o with a solution generated at random to the optimization problem to be considered. This corresponds to a state of high entropy/energy analogous to the state of solid fusion, where the particles are scattered. The initial temperature of the SA is usually determined during some initialization phase, such that almost all solution transitions will be accepted. The problem is defined in generating a set of neighborhood solutions states i and randomly choosing a state j between them. The process is accepted according to the Metropolis criterion.

This process continues until the equilibrium is found for the current temperature. The temperature is then decreased according to a predefined decreasing rule. The system is expected to reach equilibrium again and the process repeats itself. This continues until the temperature reaches a small value where no new solution is accepted. At this point the system is considered frozen and the algorithm terminates.

Typically the algorithm returns " Best So Far " (BSF) solutions, as the final solution of the algorithm. This solution is the lowest energy state through which the system has passed. You can return another type of solution, where you are (WYA - textit Where You Are), but this is unusual, since these two values are only surely equal for infinite time. A pseudo-code of the following SA algorithm is presented:

3.1.2. Mathematical Model

The SA algorithm can be modeled by Markov Chains [Geman and Geman, 1984, Hajek, 1985, Lundy and Mees, 1986, Otten and Ginneken, 1989]. The formal definition of a Markov chain is set out in Definition 2.

Definition 2. *A Markov chain is a sequence of probabilistic attempts such that the result of the current event depends solely and exclusively on the result of the previous event [Papoulis, 1991]. Markov chains are characterized by the following set of probabilities:*

$$P_r\{X_k = j | X_{k-1} = i\} = P_{ij}(k) \tag{10}$$

```
Simulated_ Annealing()
{
   // Set a initial solution i = i_o and set a initial temperature
T = T_o
   // Set a initial BSF i_BSF = i_o.
   initialize(i, t);
   i_BSF = i_o.
   do {
      do {
         // Create a new solution J with a
         // small perturbation in the solution I.
         // Then calculate the new Energy ΔE_IJ
         j=perturb(i);
         ΔE_ij = E(j) − E(i)
         // Use the Metropolis Criterion to
         // determine the acceptance of the state J
         if( ( ΔE_ij ≤ 0 ) — ( random(0,1) ) ) ¡ exp(-ΔE_ij / t) )
{
            i=j;
            // If the I state energy is smaller than the BSF state
energy
            // then BSF = i_BSF
            if ( E(i) < E(i_BSF) ) i_BSF = i
            }
         } while (not reach equilibrium)
         decrement(t)
      } while(stop criteria is not reached)
   return(i_BSF);
}
```

Figure 10. Pseudo-code for SA algorithm.

where $P_{ij}(k)$ is the probability that the k-simo event result is j, since the $(k-1)$-simo event result was i.

In the SA algorithm, the solution transitions corresponding to Markov events are clearly dependent on the previous transitions.

Extending these concepts to the SA, it can be said that $P_{ij}(t_k)$ is the probability that the result of the kth event of Metropolis is a transition from solution i to solution j at a temperature of k. This term is commonly known as transition probability. The transition probabilities on a set of solutions S at a temperature t_k form a matrix $|S|\, x\, |S|\, P(t_k)$ called the Transition Matrix. There are two factors to determine the transition probability $P_{ij}(t_k)$:

- The probability $G_{ij}(t_k)$ that a solution j will be generated after a solution i at a temperature t_k (generation probability)
- The probability $A_{ij}(t_k)$ that a solution j is accepted once it has been generated (acceptance probability)

The matrices that define the generation and acceptance probabilities are called the generation matrix $G(t_k)$ and the acceptance matrix $A(t_k)$, respectively. Thus, the transition probability $Pij(t_k)$ for each solution i in S at each temperature t_k can be defined as:

$$P_{ij}(t_k) = \begin{cases} G_{ij}(t_k) \cdot A_{ij}(t_k) & \text{for every } j \in R_i \\ 0 & \text{for the other cases} \end{cases} \tag{11}$$

The generation of the matrix G is stochastic by nature, since the probabilities of generating a neighbor next to the current solution must have a sum equal to 1. Explicitly,

$$\sum_{j \in R_i} G_{ij}(t_k) = 1 \tag{12}$$

The probabilities of acceptance are defined by the Metropolis criterion:

$$A_{ij}(t_k) = min\left[1, e^{-\Delta E_{ij}/t_k}\right] \tag{13}$$

There is also the possibility that the system remains in the i solution, where it is. In this way it is common to make explicit a probability that the model does not allow a change in the solution:

$$P_{ii}(t_k) = 1 - \sum_{j \in R_i} P_{ij}(t_k) \tag{14}$$

Substituting Eqs. 12-14, Eq. 11 can be re-written as:

$$P_{ij}(t_k) = \begin{cases} G(t_k)A(t_k) & \forall i \neq j \\ 1 - \sum_{l \in R_i} G_{il}(t_k)A_{il}(t_k), & i = j \end{cases} \tag{15}$$

As a result of this definition, we have that the transition matrix becomes stochastic in nature:

$$\forall i : \sum_{j \in R_i} P_{ij}(t_k) = 1 \tag{16}$$

If the conditional probabilities do not depend on the value of k, the Markov chain is homogeneous; otherwise, it is nonhomogeneous [Bain and Englehardt, 1992]. Algorithms based on both models were proposed in the literature. The homogeneous algorithm can be described as a sequence of homogeneous Markov chains, each executed at a fixed temperature value. The temperature decreases between subsequent chains after the current chain reaches equilibrium. The nonhomogeneous algorithm can be described as a single nonhomogeneous Markov chain. A finite number of Metropolis events is executed at each temperature possibility, and reaching equilibrium for a given temperature is no longer a criterion of convergence. [Varanelli, 1996] cites that the algorithms presented as more promising are homogeneous, given their more direct physical analogy than non-homogeneous ones.

In the next two sections, it is observed how the two algorithms are used by different authors to verify that both converge asymptotically to a set of optimal solutions S_* with probability equal to 1.

3.1.3. Homogeneous Algorithm

As discussed, the most cited articles use the algorithms for SA with the homogeneous Markov chain model. The theoretical analysis of the homogeneous models of Markov chains is presented by [Laarhoven, 1988], which shows that the algorithm converges asymptotically to the set of optimal solutions with probability equal to 1.

Unfortunately, asymptotic convergence can only be approximated in implementations of the homogeneous SA algorithm. This comes from the fact that any implementation can only generate a finite number of transitions for a finite number of temperatures. Because of this, the true balance can only be approximated by the Markov chains. In fact, no implementation can guarantee with probability 1 the arrival at a global minimum and every implementation arrives at an approximate result.

Thus, there are four parameters to fit into any homogeneous SA implementation as discussed in section 3.1. These four parameters define the implementation of the SA cooling schedule:

Definition 3. *The cooling schedule defines the exact way in which the system temperature is reduced in the course of the SA homogeneous algorithm. There are four parameters to specify for any cooling schedule:*

- *The initial temperature* t_0*;*
- *a decreasing rule to update the temperature* t_k*;*
- *a rule to determine the length* L_M *of the homogeneous Markov chain at each temperature* t_k*;*
- *a stopping criterion, specifying the conditions for terminating the execution of the algorithm.*

The cooling schedule is an attempt to approximate the theoretical algorithm in a finite computational time. It is the choice of the cooling schedule that differentiates one implementation of the SA algorithm from another. Section 3.1.4 outlines some general considerations for each parameter of the cooling schedule, as well as for the cooling schedule used in this work.

3.1.4. Cooling Schedules

As described in the definition 3, there are four parameters to be specified in the definition of the cooling schedule for each implementation of the SA homogeneous algorithm. The greatest interest in designing a cooling schedule is to achieve thermal equilibrium for the Markov chains generated at each temperature t_k. As the true equilibrium can not be achieved in the current implementations of the SA homogeneous algorithm, a way of making the system approach equilibrium (a state called a quasi-equilibrium) is necessary.

The quasi-equilibrium for a given Markov chain in a homogeneous SA algorithm is related to its stationary probability distribution. After an infinite number of transitions, the stationary distribution is the time distribution along the space of solutions. In true equilibrium, this would be the Boltzmann distribution. In quasi-equilibrium, the stationary distribution should be as close to the Boltzmann distribution as possible. This proximity measure is essential for choosing an acceptable cooling schedule.

The first choice to make is the initial temperature t_0. As the temperature approaches infinity, the stationary distribution approaches the uniform distribution over the solution space S [Lundy and Mees, 1986, Laarhoven, 1988, Otten and Ginneken, 1989,Romeo and Sangiovanni-Vincentelli, 1991]. Since the SA algorithm starts at a high temperature, the quasi-equilibrium can be quickly achieved if almost all initial transitions are accepted. This would quickly approximate the uniform distribution over the set of solutions. So a good starting point for the initial temperature is one in which almost all transitions are accepted. [Otten and Ginneken, 1989] and [Abramson et al., 1999] researched methods to obtain an initial temperature. Both authors consider generating a large number of accepted solutions, measuring the mean and standard deviation of the costs to be optimized. These values serve as an estimate of the initial cost E_∞ on the initial solution space σ_∞.

[Abramson et al., 1999] indicates that the initial temperature for the school calendar determination problem is such that:

$$t_0 = \frac{\sigma_\infty}{T} \tag{17}$$

where T is an arbitrary and high temperature. The achievement of this temperature T is done by [Laarhoven, 1988]. This method was adopted in this chapter.

The next step in building the cooling schedule is choosing the rule to reduce the temperature, or rule of decrement. When considering the decrement rule, it should be noted that large decreases will cause the system to require more Metropolis attempts to re-establish the quasi-equilibrium in the next Markov chain. [Laarhoven, 1988, Lundy and Mees, 1986, Otten and Ginneken, 1989, Romeo and Sangiovanni-Vincentelli, 1991]. As a result, there is a negotiation between longer chains and shorter decreases. These authors recommend shorter decreases to avoid freezes and very long chains.

There are essentially two types of decrement rules: fixed or adaptive. Fixed decrement rules are independent of the evolution of the algorithm in terms of aggregate statistics, such as mean and standard deviation of cost, usually in the form:

$$t_k = t_0 \cdot \alpha^k \tag{18}$$

where α is a constant such that $0 < \alpha < 1$, usually in the range of 0.90 to 0.99. As the size of the temperature decrement is kept proportionally constant over the course of the algorithm, fixed decrement rules are often called constant decrement rules. These agendas are also called classics. The following are examples of commonly cited classical rules.

Linear

The linear cooling schedule is based on the fixed subtraction of a factor β, usually proportional to the initial temperature.

$$t_k = t_{k-1} - \beta \tag{19}$$

$$\beta = t_0 \cdot 0,01 \tag{20}$$

Geometrical

This schedule is formed by multiplying a α factor, usually between $0.90 - 0.99$ [Abramson et al., 1999].

$$t_k = t_{k-1} \cdot \alpha \tag{21}$$

Adaptive schedules, on the other hand, dynamically vary the decreasing temperature over time and are commonly of the form:

$$t_k \approx t_{k-1} \cdot (\ln(\Psi_{k-1})) \tag{22}$$

where Ψ_{k-1} is a function of previous Markov chain statistics. For this reason, adaptive decrement rules are called variable rules. The following is the adaptive cooling schedule:

Geometrical Adaptative

The adaptive cooling schedule adopts a moving average in relation to the cost function to calculate the temperature decrease.

$$t_k = t_{k-1} \cdot \alpha^{k/\sigma_{k-1}} \tag{23}$$

where σ_k is the standard deviation calculated in the $k-1$ Markov chain for n attempts, usually all of the chain.

The third step in choosing a cooling schedule is to specify the stop criterion. The choice of this criterion is based on the observation that if the expectation of improvement caused by the reduction of temperature is small, the algorithm should be closed. Again, there are methods for calculating this parameter, [Abramson et al., 1999] and [Otten and Ginneken, 1989] both of which suggest the use of one of the following methods: (a) factor θ based on the standard deviation of the cost function as follows:

$$\frac{\sigma_k^2}{t_f \cdot (\mu_0 - \mu_f)} \tag{24}$$

where σ_k^2 is the variance of the cost function for the k-th Markov chain and μ_f is the average cost function for all attempts of this same string.

The last step to determine the cooling schedule is to specify the length of the L_M Markov chain. This parameter is strongly related to the size of the temperature decrease because of the quasi-equilibrium concept. Specifically, the larger the decrement, the greater the next Markov chain will need to be to re-establish the quasi-equilibrium. Since quasi-equilibrium is difficult to quantify in practice, many approaches to determining this length have been proposed. [Kirkpatrick et al., 1983] proposes that the Markov chains be closed after a number of transitions are accepted.

4. Simulated Annealing Applications to Porous Media

This topic exposes three applications of the Simulated Annealing method. The first is used to correct the grain size distribution. The other two are reconstruction methods.

4.1. Grain Size Distribution Correction

Stereology is a interdisciplinary study field which aims to evaluate the three dimensional (3D) properties of a sample using a material bi-dimensional (2D) information. Several geometric and statistical methods are applied to achieve this target [DeHoff and Rhines, 1968].

Stereology has many applications in several sciences, such as biology, medical sciences, material sciences, and wherever it is necessary to obtain information of higher dimensions from samples with inferior dimensions. The common stereology challenge is to understand the structural inner three-dimensional arrangement based on the analysis of the structure slices that show only two-dimensional information. For this reason stereological principles take into consideration geometry and probability statistics [Mandarim-De-Lacerda, 2003].

The introduction of errors during the extrapolation process is inherent to the stereological technique. [Garcia et al., 2007] state that a major problem of sampling/slicing is the loss of dimension; a 3D object becomes a 2D area, a surface (2D) becomes a line (1D), a line (1D) becomes a point (0D) and a number (points 0D) is lost. A stereological approach provides solutions to this problem.

In the middle XXth Century, stereological studies have appeared in literature with increasing frequency. The first studies are based on pioneers [Chalkley, 1943, Abercrombie, 1946, Chalkley et al., 1949, Weibel and Gomez, 1962]. The so called "new stereology" was developed in the XXth century's eighties, a collection of procedures making stereology easier and less biased [Arpat and Caers, 2007, Gundersen, 1988]. In reality, the question of bias and stereology is still under discussion, but new techniques make stereology more consistent [Hedreen, 1999, Baddeley, 2001, Bartheld, 2002, Gardella et al., 2003].

On the other hand, the idea of estimating stereological parameters from optical sections within a thick slice was first used for counting particles in optical dissector principle [Sterio, 1984] and in unbiased sampling brick rule [Howard et al., 1985], and then in many other stereological methods. Methods examples are: nucleator [Gundersen, 1988] and planar rotator [Jensen and Gundersen, 1993] applied to a stack of optical sections and estimating the mean particle volume, spatial grid [Talukdar and Torsaeter, 2002], methods for estimating the surface area, method of vertical slices [Gokhale, 1990], and global spatial sampling [Larsen and Nielsen, 1998] used for the length estimation.

Indeed, stereology is a dynamic science and new approaches to old questions still stimulate stereologists to test possibilities, an exciting example is the Simulated Annealing Technique for curve fitting.

This research evaluated several functions as candidates to represent 3D particle size distributions which generated 2D particle size distributions experimentally measured through cut porous media samples and image analysis. The evaluation determined the most adequate function for representing the particle size distribution and a new alternative was also proposed: not using a specific function, but to optimize a constant distribution.

4.2. Particle Size Distribution Modeling

Some considerations were necessary to model a particle size distribution:

- each particle is represented by a sphere;

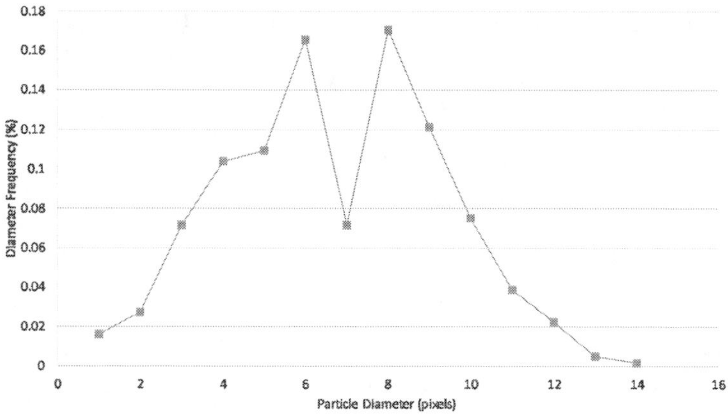

Figure 11. Typical particle size distribution for a porous media obtained by image analysis.

- the spheres are non-penetrating;
- the 3D particle size distribution follows a known unimodal distribution, as a log-logistic function with unknown parameters which will be optimized by a Simulated Annealing method;

These considerations aim to make the sample particles model as real as possible, but yet feasible to optimize. The unimodal function consideration intends to simplify the optimization procedure. Even if the Figure 11 can be considered a bimodal function, it is considered that the oscillation around the 30μm diameter is a fluctuation.

The following functions were chosen to model the particle size distribution:

$$f(x,\mu,\sigma) = \frac{1}{x\sigma\sqrt{2\pi}} e^{-\frac{(ln(x)-\mu)^2}{2\sigma^2}} \tag{25}$$

$$f(x,\mu,\beta) = \frac{e^{-\frac{x-\mu}{\beta}} - e^{\frac{x-\mu}{\beta}}}{\beta} \tag{26}$$

$$f(x,\beta,n,\delta) = \beta e^{-\beta(x+\delta)} e^{-ne^{-\beta(x+\delta)}} \{1 + n[-e^{\beta(x+\delta)}]\} \tag{27}$$

$$f(x,\alpha,\beta) = \frac{(\beta/\alpha)(x/\alpha)^{\beta-1}}{[1+(x/\alpha)^{\beta}]^2} \tag{28}$$

These functions, Log-Normal - 25, FischerTippet - 26, Shifted-Gompertz - 27 and Log-Logistic - 28, were chosen as unimodal functions and due to its shape representation possibilities. The Log-normal function, for example in Figure 12 can have several shapes. The Figure shows The Log-Normal function for $\mu = 0$ and different values of σ.

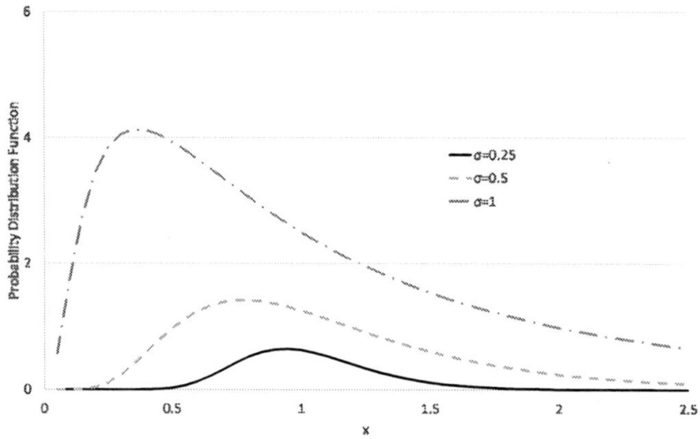

Figure 12. Log-Normal function shapes varying the σ parameter.

For all these functions the function parameters were randomized and a 3D initial particle size distribution was generated. The 2D particle size distribution was generated through virtual sphere slices. For each distribution radius, a number of 105 times the distribution spheres was generated. These spheres are then cut and 2D particles (circles) with smaller radiuses are then created, as shown in Figure 13. R is the real radius, r, the measured radius and h is the distance from the center where the slicing plane cut the particle. For example, a particle size distribution which has 0.01 (1%) of its particles of radius 3, will have, for this radius, 1.000 spheres generated, which will be virtually cut and these cuts can create 2D particles measurements with radius 3, 2 and 1.

Both the 2D particle size distribution created by a 3D particle size distribution can be observed in the Figure 14 for a Log-Normal distribution. As expected, the created 2D curve has a higher frequency of particles with smaller diameter than the 3D curve.

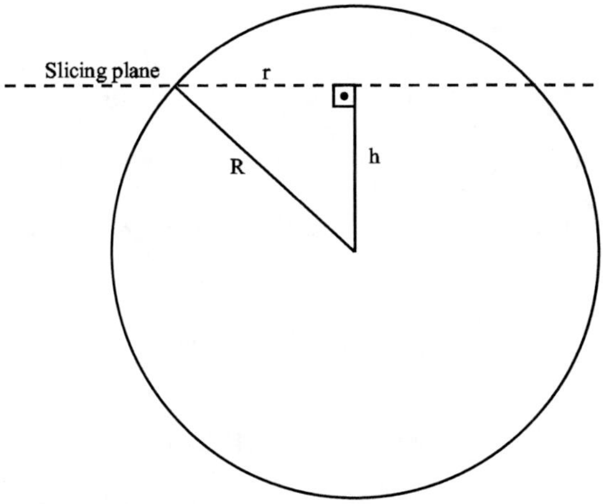

Figure 13. Sphere (Particle) cut by a slicing plane illustrating the measurement of a r radius, smaller than the true radius R.

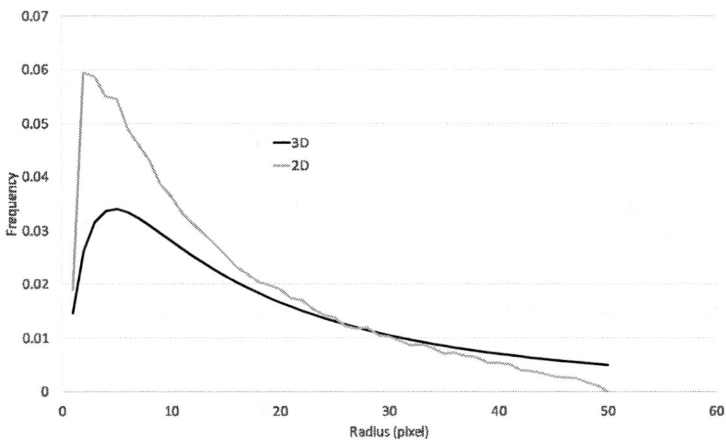

Figure 14. Log-normal curve and the created 2D after the virtual cuts.

4.2.1. Simulated Annealing Curve Optimization

The well-known SA technique was used to optimize the 3D particle size distribution. Once the 3D particle size distribution was generated, a SA algorithm was used to adjust the curve parameters to fit the 2D grain size distribution. The sequence of steps to optimize the curve is described below:

1. Read the grain size distribution 2D;
2. Choose the equation to represent the distribution;
3. Generate random parameters for the chosen equation;
4. Generate a large number (around 10,000) of spheres according to the chosen equation;
5. Perform a virtual cut in all spheres and generate a 2D grain size distribution;
6. Compare the read and generated 2D distributions;
7. Modify a parameter of the equation through an SA algorithm;
8. Change the parameter to be modified by the SA;
9. Repeat steps 4 to 8 until the desired error is reached or the energy freezes;
10. Save the parameters of the equation, reduce the resolution of the parameters and return to step 4;
11. Quit the program with a 3D grain size distribution output.

It is remarkable that the optimization is carried on two levels. While the 3D curve is directly modified, the 2D generated curve provides the square error. This configuration is due to the experimental procedure, which provides a 2D particle size distribution, so it is necessary to compare both 2D distributions. The square error was calculated as equation 29:

$$E = [\sum f(r)_{generated} - f(r)_{2D-experimental}]^2 \tag{29}$$

The SA algorithm used in this research is the classical algorithm with linear cooling schedule, having $T_{c+1} = 0.99 \cdot T_c$. The cooling happened after 1000 iterations with reducing energy.

The parameter modification was executed at one parameter per iteration. For the Log-Normal function, for example, if the t iteration modified the μ parameter, the t+1 iteration modified the σ parameter. The modification step was 0.01 and it could be added or subtracted with a 50% chance for each option. The convergence target was 10^-3 for the square error. The iteration limit was 10.000 iterations without improvement in the energy function (square error).

A typical result for the same particle size distribution from Figure 11 can be observed in Figure 15. The square error for this result was 9x10-4.

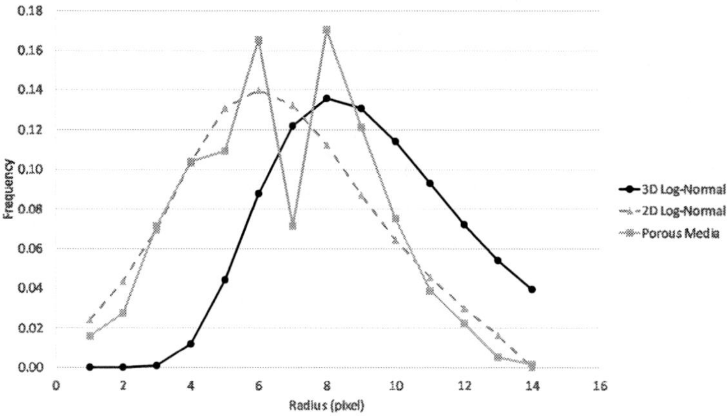

Figure 15. Processing result for the particle size distribution presented in Figure 11.

To test which function could better model the particle size distribution, a SA process was executed for 21 particle size distributions for all the adopted functions without ending the processing with the convergence criteria. An average square error for all the samples can be observed in the Table 1. The function with the best result was the Log-Logistic. The square error obtained was $7.63E^{-}04$. This indicates that the best function for representing the particle size distribution among the analyzed functions is the Log-Logistic.

Table 1. Average square error obtained for unimodal functions

Function	Average square error obtained
Log-Normal	$1.03E^{-03}$
Shifted-Gomperz	$8.56E^{-04}$
Fisher-Tippett	$8.14E^{-04}$
Log-Logistic	$7.63E^{-04}$

To compare the proposed methodology with classic methods, a Log-Normal distribution with 10 particle diameters, μ=1.3 and σ=0.4 was adopted as a 3D particle size distribution and the virtual cut was executed. The resultant distributions are shown in Figure 16.

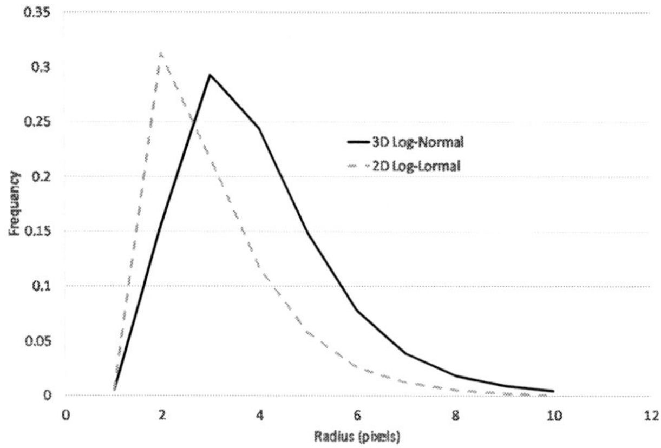

Figure 16. Log-Normal curve with 10 particle diameters, μ=1.3 and σ=0.4 and 2D curve resultant from virtual cuts.

The classical method of Schwarz-Saltykov [DeHoff and Rhines, 1968], which could be adapted for a curve correction, was applied to the resultant 2D curve and a 3D corrected particle size distribution was generated. The latter was compared with a particle size distribution generated using the proposed methodology through quadratic error calculated in the 3D distribution. The corrected distributions and the resulting curves are disposed in Figure 17. The quadratic error for the Schwarz-Saltikov method was $2.2E^{-1}$, while the SA function was $8E^{-2}$, approximately 5 times smaller error.

It is remarkable that even if the error was almost 5 times smaller, the 3D distribution was not satisfactory. The SA function was not able to represent a Log-Normal curve with a Log-Logistic curve, even if both distributions are theoretically compatible.

A proposed solution is not to use a function to represent the 3D particle size distribution. Instead a constant function or a linear function can be used and let the SA algorithm modify it freely. A preliminary result of a constant modifying function with modifying step $1E^{-3}$ is shown in Figure 18. The 2D input was the same 2D Log-Normal as the previous example and the 3D result is compared with the previous Log-Normal.

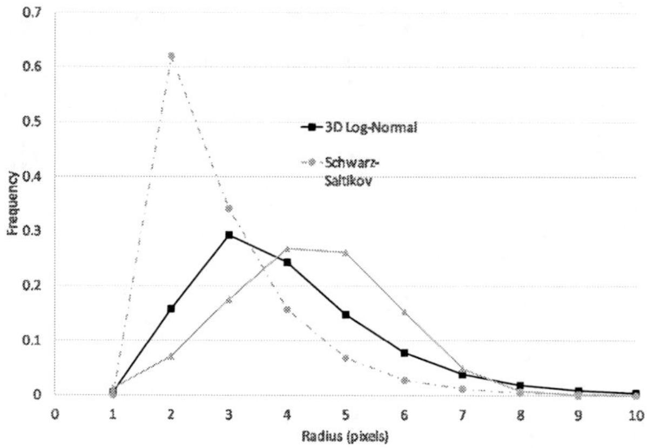

Figure 17. Comparison between 3D particle size distributions generated by the proposed methodology and the classical Schwarz-Saltikov method.

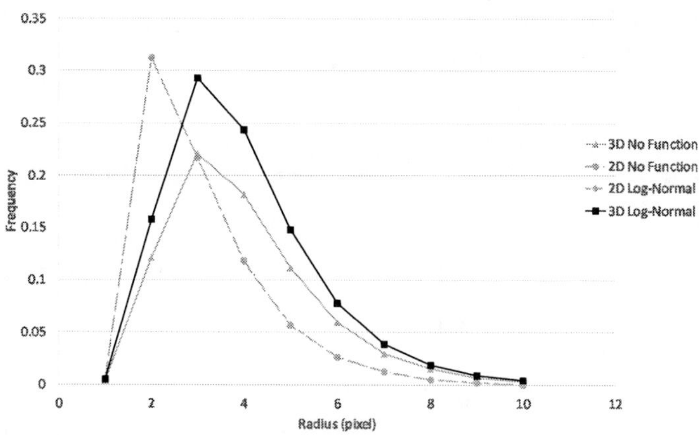

Figure 18. Comparison between 3D particle size distribution: generated by the SA-no function and previous Log-Normal function.

There is a significant agreement between both 2D functions (it is not possible to differentiate both visually). The quadratic error for this methodology was $1.2E^{-3}$, even smaller than the SA with a function representation. It can also be noticed that the 3D functions did not match completely, however the result is better than the previous methods.

This methodology was also tested with the grain size distribution showed in Figure 11. The resulting curves are presented in Figure 19.

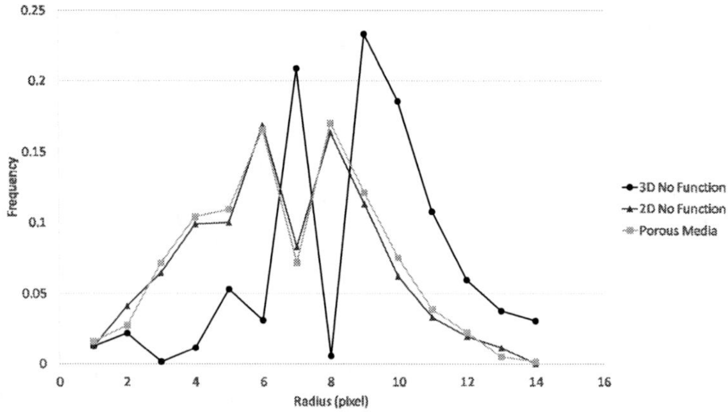

Figure 19. 3D Curve generated using a SA technique without optimizing a specific function and its 2D cut compared to a sandstone grain size distribution.

An interesting fact is that the new methodology was able to reproduce a bimodal function and the 3D generated curve was also bimodal.

5. Porous Media Reconstruction Using Simulated Annealing Techniques

The problem of reconstructing a porous media from two to three dimensions is stochastic and combinatorial by definition. A binary image of a porous microstructure is nothing more than a set of white and black pixels arranged in an organized way. Given the nature of this, the SA method can be used to reconstruct 3D porous media retaining certain properties of 2D images obtained from optical images. Several authors [Ouenes and Bhagavan, 1994, Rintoul and Torquato, 1997, Torquato and Kim, 1999, Biswal et al., 1998, Cule and Torquato, 1999, Manwart et al., 2000, Hilfer, 2000, Sheenan and Torquato, 2001, Torquato, 2002, Talukdar and Torsaeter, 2002] have proposed parameters to be used as cost functions, models of neighborhoods besides Euclidean, and optimizations in the method to enable the use of SA for the reconstruction problem. These authors have succeeded in replicating parameters such as the autocorrelation function and the 2D chord size distribution, but [Okabe and Blunt, 2004] exposes problems of connectivity and pore size when trying to reconstruct microstructures in 3D.

In general, the reconstruction algorithms using SA have been based on two types of movement of pixels or voxels:

- Exchange of black and white pixels' coordinates, starting from an image whose original porosity is the same as the microstructure that it is desired to reconstruct;
- Exchanging values of white pixels for black and vice versa, starting from a random image and allowing the system to organize the porosity according to the cost function.

The two approaches are based on moving pixels. There is also another class of movement called "moving objects". This is based on moving sets of pixels instead of pixels in isolation. [Rintoul and Torquato, 1997] used this method when moving circles of a given radius in a 2D image in order to reproduce a second order function called the radial distribution function. [Stoicaa et al., 2005] used SA with ball movement as a tool for pattern analysis.

Two methods for reconstructing 3D media from 2D data are presented in this chapter: A method based on moving pixels while retaining the autocorrelation function, DBI distribution, and chord length distribution: Pixel- Based Simulated Annealing (PSA) and Object-Based Simulation Annealing (OSA). In both cases, the quality of a reconstruction is measured by the energy (quadratic error) of the system for the optimized parameters.

5.1. *Pixel-Based Simulated Annealing* (PSA)

This method is based on the isolated movement of pixels in order to preserve one or more functions obtained from 2D images. Most authors who worked with SA for the reconstruction of porous media used this method while retaining the autocorrelation function or functions of the same family, such as the pore-solid correlation [Manwart et al., 2000]. [Sheenan and Torquato, 2001] also used a hexagonal pixels neighborhood to better attempt to represent the porous medium. It is known, however, that using only the autocorrelation function is not sufficient to maintain connectivity in an image [Okabe and Blunt, 2004]. Thus, it was necessary to include other parameters in the cost function of

SA. [Talukdar and Torsaeter, 2002] has been relatively successful in reproducing 2D images of glass beads using the functions derived from autocorrelation and linear chord (this is to increase connectivity) defined by [Torquato, 2002]. However, it is possible to observe the presence of isolated pixels in the pore phase in the reconstructions resulting from their methodology. This is observed in the comparison between Figure 20 (original) and Figure 21 (reconstruction), extracted from [Talukdar and Torsaeter, 2002].

Figure 20. Original image of an aggregate of glass beads with black porous space.

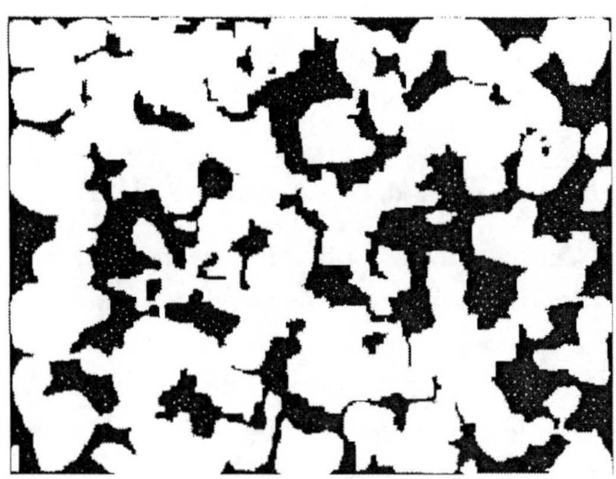

Figure 21. Reconstructed image using SA optimizing the functions derived from autocorrelation and linear chord with error $E = 10^{-4}$.

Given the parameters used by the cited authors, it was decided to include a third parameter to be optimized by the PSA: the DBI distribution. This parameter aims to eliminate white pixels that remained isolated in the pore phase even after optimizing the functions used until that point. Thus, the PSA has as cost function the sum of three parameters: the autocorrelation function, the chord size distribution (normalization of all linear chord sizes) and the DBI distribution.

For all simulations using this method the geometric cooling schedule was adopted with $\alpha = 0.999$. This configuration was chosen given its slow and steady decrease over the interactions. The temperature was reduced every 500 iterations without improvement in the energy value, that is, the homogeneous algorithm was adopted. The algorithm returns BSF type solutions. Reheating schedules have not been used, because of the large space of solutions, the small influence of the movement of a pixel or voxel in the square error, the system tends to not get trapped in local minimums.

It can be seen in Figures 23 and 26 the operation of the PSA for two synthetic samples shown in Figures 22 and 25, respectively. It starts processing with a random image of the same porosity, as shown in Figure 24 and moved the pixels until the desired result is reached, that is, the quadratic error of the functions optimized to be less than 10^{-5}. The average time for a 2D reconstruction of a 400x400 pixel image was 24 hours using a 1GHz processor.

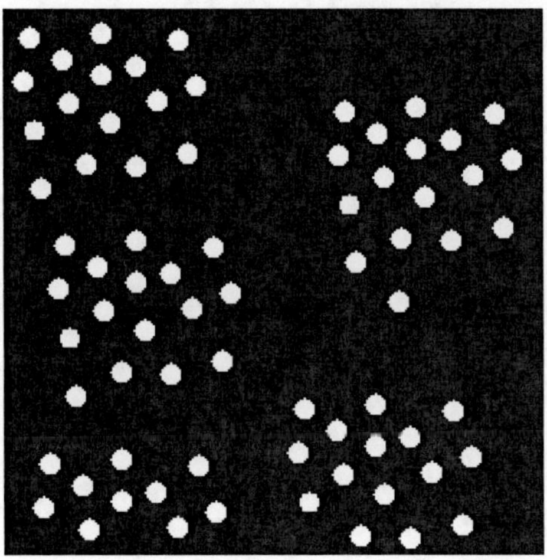

Figure 22. Original image of circles.

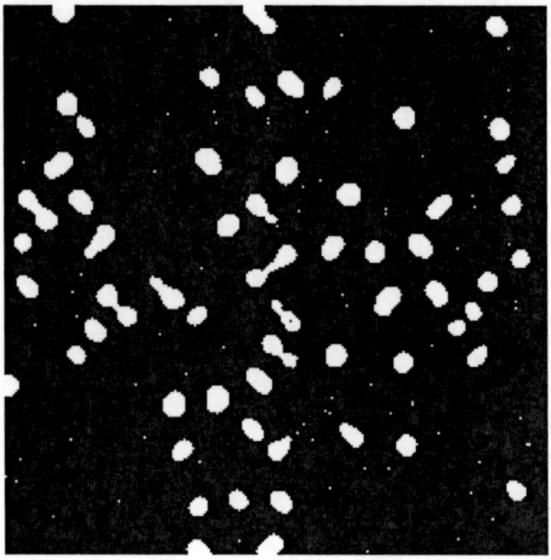

Figure 23. Image of circles reconstructed using the PSA method using the parameters autocorrelation function, chord size distribution and distribution of DBI.

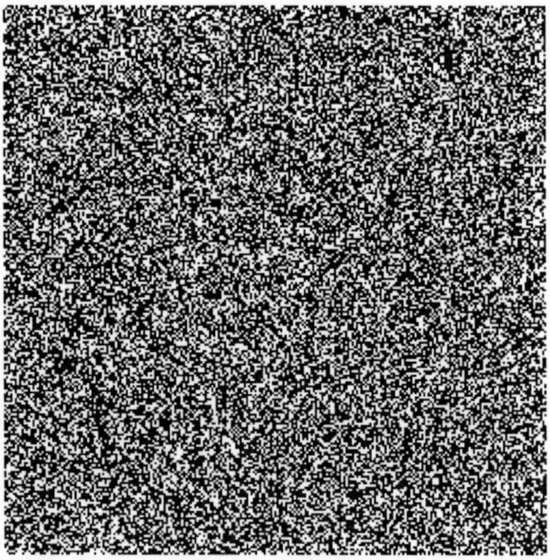

Figure 24. Initial image for the processing of Figure 25.

A reduction in the number of isolated pixels is observed when the DBI distribution parameter is used, as expected, but the existence these pixels is yet noticed. Considering the reduction of these, it was accepted as an error. Using

Figure 25. Original image of squares.

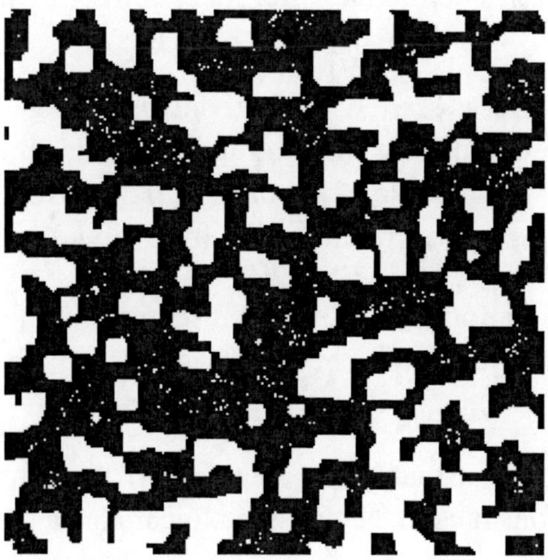

Figure 26. Image of squares reconstructed using the PSA method using the parameters autocorrelation function, chord size distribution and DBI distribution.

a median filter of size 1 in Figure 23, it can be seen in Figure 27 that the deletion of the isolated pixels did not significantly influenced the autocorrelation function. The square error alteration was $1.59E^{-5}$.

Once the problem of the isolated pixels was worked around, the 3D re-

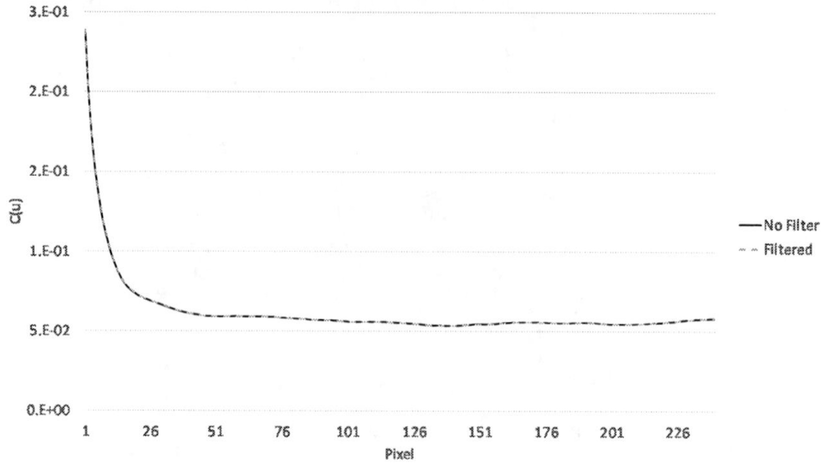

Figure 27. Comparison between the autocorrelation function of a reconstructed image of the circles before and after median filter of size 1.

construction step was started. The reconstruction used the same methodology, starting from a completely random image of the same porosity of the desired sample (average porosity of 2D images). The voxels (3D pixels) were moved from the 3D image while retaining the following functions:

- Autocorrelation per section - All sections of the 3D image in the 3 orientations (vertical, horizontal and sagittal) must have the same autocorrelation.
- Chord size distribution per section - All sections of the 3D image in the 3 orientations must have the same linear chord function.
- DBI distribution per section - All sections of the 3D image in the 3 orientations must have the same DBI distribution.

These parameters should guarantee a completely homogeneous and isotropic image. It is observed that three functions were used to optimize, considering the original images as isotropic and homogeneous. This consideration can be reviewed and the same functions can be used independently of the guidelines, totaling up to 9 functions and generating non-homogeneities and/or anisotropy.

For all 3D simulations using this method it was adopted the geometric cooling schedule with $\alpha = 0.98$. This schedule decreases faster than that adopted in the PSA method. This change was chosen given a smaller space of solutions that the ball movement, instead of pixels, provides. The temperature was reduced every 5,000 iterations without improvement in the energy value. Note that although the temperature decreases faster, more interactions are required so that each Markov chain reaches convergence and the temperature can be reduced. The algorithm also returns BSF type solutions. The program terminates when the desired error criterion is reached ($E <= 10^{-6}$) or the temperature has already been reduced enough not to find energy reductions per 10,000 interactions. The average processing time was 173h for an image of 80^3 voxels.

It can be seen the result of a 3D reconstruction of Figure 25 in Figure 28, with the solid phase in blue and gray. It should be observed the objects with sharp corners, indicating the adequate representation of the squares. This image had a quadratic error of $2.3 \cdot 10^{-5}$ with respect to the autocorrelation of all sections.

Figure 28. 3D reconstruction with 80^3 voxels of Figure 25 through the PSA method.

5.2. *Object-Based Simulated Annealing* (OSA)

The basic difference of this method for the PSA lies in the movement of its neighborhood. Instead of moving pixels or voxels, move circles (for 2D) or spheres (for 3D). The spheres are generated by measuring the grain size distribution of the sample 2D images to be reconstructed. Since the grain size distribution is conserved by the grain generation itself, and the connectivity is preserved by the very nature of the method, it is not necessary to optimize the function of the chord size distribution; only the autocorrelation.

For the 2D reconstructions the same cooling schedule of the PSA method was adopted, but the temperature reductions were made every 50 interactions without energy reduction. This was chosen because of the difference in the stage of neighborhood movement. The OSA method converges more quickly to a desired minimum. As expected, the processing time was much lower for this method than for PSA. For a 400x400 pixel image, the average time for a reconstruction is 8h.

One can observe, respectively, in Figures 29 and 30 the initial and final states of a reconstruction of a porous media sample. Note that the reconstruction style is quite different from the PSA method. The initial image does not have the same porosity of the final image, since the circles can overlap and change this parameter, however, with the optimization of the autocorrelation function, the final image will have the desired porosity. The figure 31 shows the evolution of the autocorrelation function for the previous images.

For a 3D reconstruction there is a point to deal with. The 2D grain size distribution used to generate the circles in the 2D reconstructions is not equal to the one that should be used to generate 3D spheres. This occurred to the sectioning process of the sample through the 3D grain size distribution as shown in Figure 13. The 2D distribution tends to overestimate the grains of smaller diameters and underestimate the larger diameter grains. In this figure, the diameter of the grain is R, whereas the one measured after the sectioning will be r.

This problem can be approached applying the methodology exposed in section 4.1. The 2D grain size distribution was corrected to a 3D grain size distribution. This approach brings several benefits to the OSA method, such as the reduction of the small grains frequency, which makes the convergence faster.

After a 3D distribution was generated, the Equation 30 was used to calculate the number of spheres for each diameter of the grain size distribution.

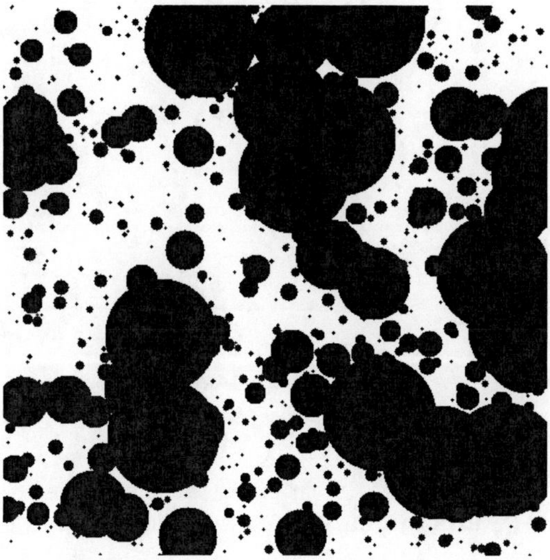

Figure 29. Initial image for a reconstruction of a porous media (porosity=41.1896%).

Figure 30. Final image for a reconstruction of a porous media (porosidade=15,8244%).

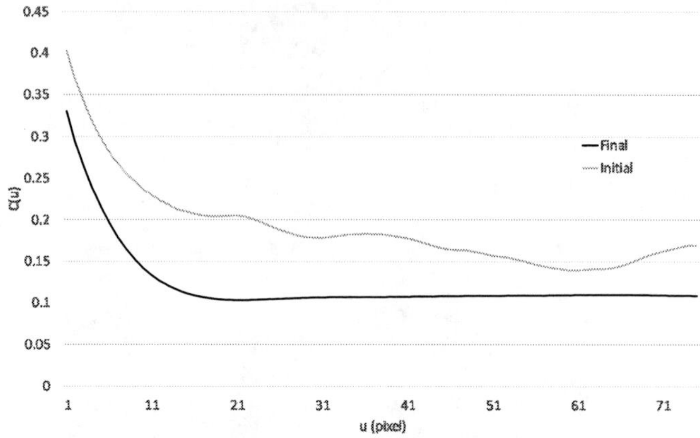

Figure 31. Autocorrelation function measured in Figures 29 and 30.

$$n(r) = \left(\frac{3}{4}\right)^{d-2} \cdot \frac{L^d}{\pi r^d} \tag{30}$$

Where d is the dimension (2 or 3), L is the linear size of the reconstructed image, and r is the radius of each element distribution of the grain size distribution.

On the other SA parameters, the same cooling schedule of the PSA method was adopted, but the temperature reductions were performed every 100 interations without energy reduction. Like the 2D algorithm, the system converged faster than the PSA method.

It can be seen in Figure 32 the result of a reconstruction of a porous media with the solid phase in blue. Like the 2D case, the porosity of the sample varies throughout the processing, but the final porosity is the same as desired. This image had a quadratic error of $8.7 \cdot 10^{-5}$ with respect to the autocorrelation of all sections.

6. Results and Discussion

This section presents the results obtained from 2D and 3D reconstructions of porous media. To validate the reconstructions, comparisons of geometric parameters (correlation and pore size distribution) were performed with other reconstruction methods and their microtomographic images.

Figure 32. 3D reconstruction with 100^3 Voxels of a porous media using the OSA method.

6.1. 2D Results

In this topic, the 2D geometric parameters measured in thin-slice sections as well as the stochastic ones of the same images using Pixel-based Simulated Annealing (PSA) and Object-based Simulated Annealing (OSA). Ten samples of porous media were reconstructed using the PSA, OSA, Truncated Gaussian (TG) and Superposed Spheres (SS) methods and their reconstructions were compared by measuring the quadratic error in relation to the correlation and the pore size distribution of the original image.

6.1.1. Correlation Function

Table 2 exposes the porosities and spatial resolutions for each sample, while the Table 3 shows the results for the quadratic error regarding the correlation function. For the TG method, four amplification factors were reconstructed in order to verify which one better reproduced the original microstructure. The number presented in parentheses indicates the amplification factor that presented the smallest error. This methodology was adopted for all the results presented in this section. All images had dimensions of 400x400 pixels. The SS was not compared due to its poor correlation function reproduction.

It is observed that the OSA method obtained better results for 9 of the 10 samples, whose porosity ranged from 13.80 % to 30.37 %. The PSA method did not perform well in replicating the correlation for the 2D samples. This can

Table 2. Porosity and spatial resolution for 10 porous media samples

Sample	Porosity	Spatial Resolution
01	13.24%	7.84μm
02	13.80%	7.84μm
03	15.66%	7.84μm
04	18.84%	7.84μm
05	19.50%	3.12μm
06	20.84%	7.84μm
07	23.93%	7.84μm
08	26.81%	3.12μm
09	30.37%	3.87μm
10	30.78%	2.03μm

Table 3. Quadratic error regarding the Correlation function for 10 samples of porous media reconstructions using the PSA, OSA and TG methods

	Error regarding the Correlation function			
Sample	**PSA**	**OSA**	**TG-Best**	**Minimal**
01	$1.85{\cdot}10^{-3}$	$4.75{\cdot}10^{-3}$	$1.80{\cdot}10^{-3}$ (2)	TG-F2
02	$7.74{\cdot}10^{-4}$	$6.08{\cdot}10^{-6}$	$4.10{\cdot}10^{-5}$ (3)	OSA
03	$1.89{\cdot}10^{-3}$	$6.56{\cdot}10^{-5}$	$1.16{\cdot}10^{-4}$ (4)	OSA
04	$5.98{\cdot}10^{-4}$	$1.84{\cdot}10^{-5}$	$1.74{\cdot}10^{-4}$ (4)	OSA
05	$1.53{\cdot}10^{-3}$	$2.12{\cdot}10^{-5}$	$1.67{\cdot}10^{-3}$ (4)	OSA
06	$1.28{\cdot}10^{-3}$	$7.85{\cdot}10^{-5}$	$4.11{\cdot}10^{-4}$ (4)	OSA
07	$2.93{\cdot}10^{-3}$	$2.45{\cdot}10^{-5}$	$3.57{\cdot}10^{-4}$ (2)	OSA
08	$2.99{\cdot}10^{-2}$	$2.82{\cdot}10^{-5}$	$4.27{\cdot}10^{-4}$ (3)	OSA
09	$4.03{\cdot}10^{-4}$	$6.08{\cdot}10^{-5}$	$1.72{\cdot}10^{-3}$ (3)	OSA
10	$7.75{\cdot}10^{-4}$	$4.54{\cdot}10^{-4}$	$7.35{\cdot}10^{-4}$ (1)	OSA

occur due to insufficient processing time. It would thus be necessary to increase the processing time through a slower reduction of the temperature parameter. The mean time of reconstruction using the PSA method was 16h, while the

OSA was 20min.

In Figure 33 it can be seen the original binary image of Sample 01, obtained through thin slides. For each sample, 10 similar images were used to measure the desired parameters to be used in the reconstruction: correlation and chord size distribution and DBI. The mean of the measurements of the parameters of all the images was used as reference.

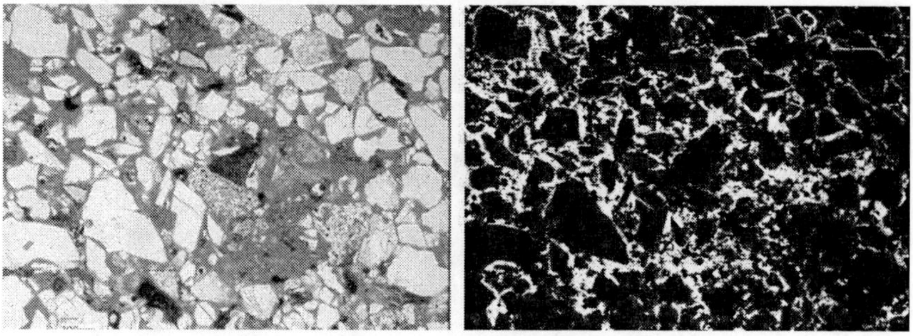

Figure 33. Flat-section image of Sample 01 with 7.84 μ m resolution, as well as its respective segmented image with 13.24 % porosity.

In Figures 34 and 35 it is showed the results of the 2D reconstructions for Sample 01. Note that the reconstruction of Figure 34 presented isolated pixels, despite the parameters used to prevent this from happening. The presence of the same ones in smaller quantity in the original image is observed. These were measured, especially in the linear chord distribution and DBI distribution functions, and this noise was reproduced in the reconstructions. This did not occur with the OSA method, since this method does not take these parameters into account, but the connectivity information is reproduced through the movement of the objects. Figure 36 presents the reconstruction result for the TG method, which obtained the best result for this case. It is noted that although the correlation function was the least quadratic error, the pores formed are of small diameters, much of it, isolated pixels.

6.1.2. 2D Pore Size Distribution

Table 4 shows the results of the same reconstructions for the pore size distribution with the metric d_{3-4}. The results are compared with the SS method.

It is also noted in Figures 37 and 38, the graph of the pore size distribution for Sample 05. The TG method formed excessive small pores so that was placed

Figure 34. Sample 01 reconstruction image result using PSA method 400x400 pixels and 13.24% of porosity.

Table 4. Comparison of the Pore Size Distribution between the reconstruction of 10 samples of porous media using the PSA, OSA, TG and SS methods

	Error regarding the d_{3-4} Pore size distribution				
Sample	**PSA**	**OSA**	**TG-Best**	**ES**	**Minimal**
01	$1.34{\cdot}10^{-3}$	$3.19{\cdot}10^{-2}$	$2.38{\cdot}10^{-1}$ (3)	$6.82{\cdot}10^{-3}$	PSA
02	$1.97{\cdot}10^{-3}$	$5.73{\cdot}10^{-3}$	$1.57{\cdot}10^{-1}$ (2)	$4.42{\cdot}10^{-3}$	PSA
03	$1.28{\cdot}10^{-2}$	$1.27{\cdot}10^{-2}$	$1.60{\cdot}10^{-1}$ (4)	$2.83{\cdot}10^{-2}$	OSA
04	$1.95{\cdot}10^{-3}$	$1.27{\cdot}10^{-2}$	$1.23{\cdot}10^{-1}$ (3)	$5.80{\cdot}10^{-3}$	PSA
05	$8.24{\cdot}10^{-3}$	$5.92{\cdot}10^{-3}$	$2.13{\cdot}10^{-1}$ (4)	$1.47{\cdot}10^{-2}$	OSA
06	$2.68{\cdot}10^{-1}$	$5.07{\cdot}10^{-3}$	$2.05{\cdot}10^{-1}$ (3)	$4.91{\cdot}10^{-3}$	SS
07	$2.05{\cdot}10^{-3}$	$1.51{\cdot}10^{-3}$	$5.60{\cdot}10^{-2}$ (2)	$1.25{\cdot}10^{-2}$	OSA
08	$1.53{\cdot}10^{-2}$	$6.79{\cdot}10^{-3}$	$6.39{\cdot}10^{-2}$ (3)	$7.01{\cdot}10^{-3}$	OSA
09	$7.42{\cdot}10^{-3}$	$2.99{\cdot}10^{-3}$	$8.63{\cdot}10^{-3}$ (1)	$5.67{\cdot}10^{-3}$	OSA
10	$3.26{\cdot}10^{-3}$	$2.43{\cdot}10^{-3}$	$4.03{\cdot}10^{-3}$ (1)	$7.73{\cdot}10^{-3}$	OSA

separately from the other methods for better visualization of these in comparison with the original distribution. The PSA and SS methods formed pores of smaller size than those of the OSA and those of the original image. The OSA, despite

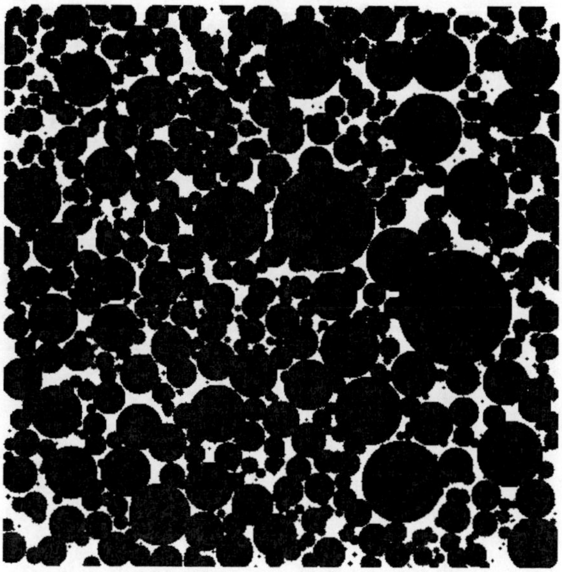

Figure 35. Sample 01 reconstruction image result using OSA method 400x400 pixels and 13.24% of porosity.

not achieving pores of all sizes, was the method with the best performance. It can be conclude that, for these samples, the reconstruction method that best reproduced this parameter was the OSA.

The other samples presented similar results, differing only in sample 01, 02 and 04, where the PSA method had a better performance. The grain size distribution of Sample 01 would require a longer processing time for the OSA method, effected through a slower reduction of the temperature parameter. In this way, the system is considered "frozen" in a local minimum. However, it was decided not to change the parameters and accept this simulation as being unsuccessful in relation to the PSA and TG methods.

It is observed that the OSA and PSA methods obtained better results than the TG and ES methods for 9 cases of the 10 and the OSA was the best method for 6 samples. This result was somewhat surprising, since the PSA method retains the DBI distribution, so it would be likely to yield better results than the OSA method. This result indicates that the fact that the OSA method retains only the correlation function does not significantly influence the conservation of the pore size distribution.

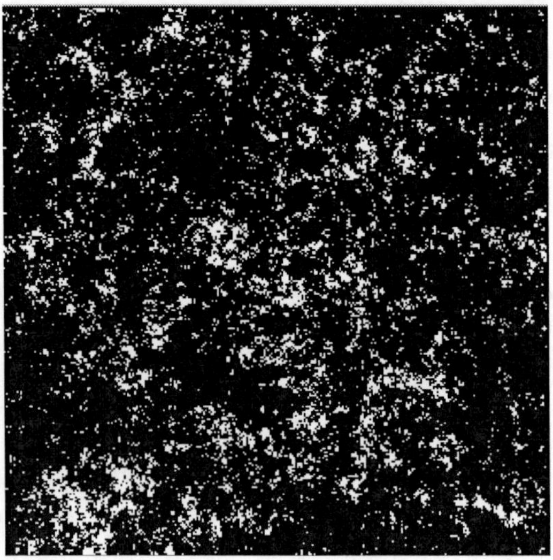

Figure 36. Sample 01 reconstruction image result using TG(2) method 400x400 pixels and 13.24% of porosity.

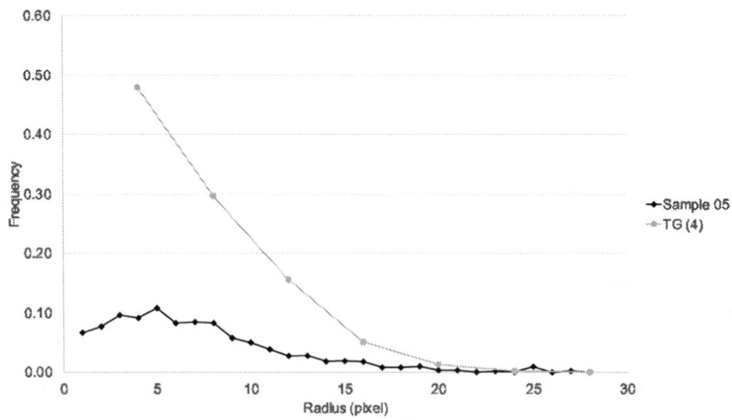

Figure 37. Comparison between the graphs of the d_{3-4} pore size distributions measured in Sample 5 and its reconstruction using the TG.

The PSA method, despite preserving the DBI distribution function, was not better than the OSA method. This was attributed to insufficient processing time, but given the better performance of the OSA method with a much shorter pro-

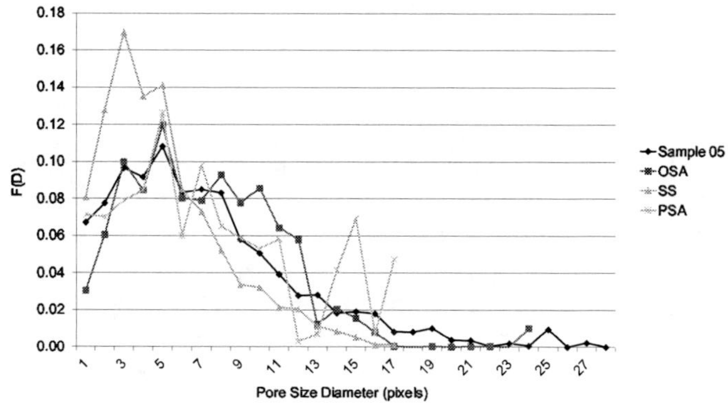

Figure 38. Comparison between the graphs of the d_{3-4} pore size distributions measured in Sample 5 and their reconstructions.

cessing time, it was decided not to change the parameters used.

It was concluded, therefore, that the OSA method was presented as the most suitable to reconstruct the samples used in this work in 2D.

6.2. 3D Results

For the 3D case, the same parameters were analyzed. The reference, however, was changed from 2D images to microtomographic images. In this chapter, the microtomographic images were used to make measurements of geometric parameters, in order to validate the reconstruction methods, which are used when they are not possessed. Thus, it was possible to analyze the 3D pore size distribution calculated through the metric d_{3-4-5}. It was also analyzed the geometric parameters for the 2D sections of the reconstructed images and the microtomographies, comparing the mean square error in relation to the correlation function.

It was used 5 3D images of samples of porous medium whose number of acquired sections for each sample, spatial resolutions and tensions and currents applied in the tube for acquisition of the images are exposed in the Table 5.

The samples were approximately cylindrical, 2 mm diameter and 4 mm high. A typical profile of the porosity along sections for Sample 15 can be seen in Figure 39, while the microtomographic image for the same sample can be seen in Figure 40. The Sample 15 was purchased on Skyscan 1072 equip-

Table 5. Number of acquired sections, spatial resolution and acquisition parameters for 5 porous media microtomographic images

Sample	N°of sections	Spatial resolution	Tension/Current applied in tube
11	804	5μm	80kV / 124μA
12	831	2μm	50kV / 200μA
13	979	5μm	80kV / 124μA
14	804	5μm	80kV / 124μA
15	956	3.81μm	60kV / 165μA

ment and presented by [Appoloni et al., 2005]. This equipment is similar to model 1172 used for the acquisition of other samples.

Figure 39. Porosity profile of a sample used in this chapter, with emphasis on the porosity measured using image analysis.

A two-dimensional section of this sample and the cutout used for the study of this chapter are presented in Figure 41.

To perform the analyzes it was necessary to binarize the images through the gray level threshold. Table 6 shows the porosity measured for each sample and the cubic volume in voxels used for the analysis in this work. For each sample, a representative cubic volume (greater than twice the correlation range) was selected. For Sample 15, only 150 images were accessed, although the acquisition was 956 images, but as its correlation range measured 40, the volume could be considered representative, and consequently, used in the present chapter.

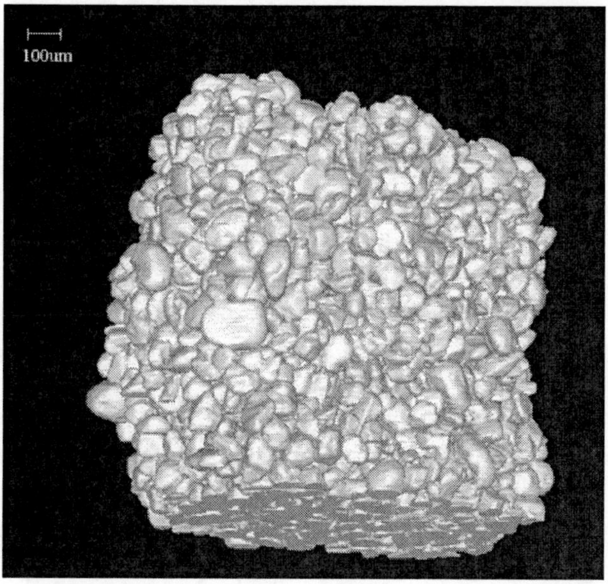

Figure 40. Microtomographic image of a porous media presented by [Appoloni et al., 2005] (Sample 15).

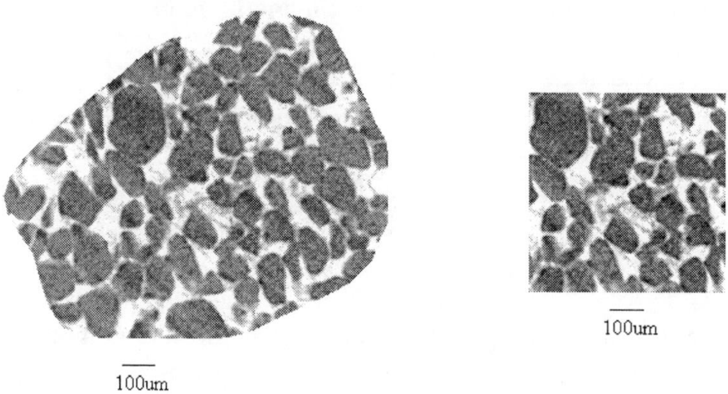

Figure 41. Two-dimensional microtomographic section of a porous media (Sample 15) presented by [Appoloni et al., 2005] and its respective cut-off used in this chapter.

Table 6. Porosity, used volume and correlation range for each sample

Sample	Porosity	Used volume (voxels)	Correlation range (voxels)
11	9.30%	250^3	60
12	14.64%	250^3	60
13	14.68%	450^3	70
14	28.14%	200^3	40
15	33.04%	150^3	40

6.2.1. Correlation Function

For the reconstruction of the same samples PSA, OSA, OSA with grain size distribution correction (OSA*), TG and ES were used and their reconstructions were compared by measuring the quadratic error in relation to the correlation and pore size distribution. Table 7 shows the results for the calculated error. The reconstructions had 200^3 voxels, with the exception of Sample 15, whose reconstructions had 150^3 voxels.

Table 7. Comparison of the correlation between the reconstruction of 5 samples of porous media using the methods PSA, OSA and TG

	Errors regarding the correlation function				
Sample	PSA	OSA	OSA*	TG-Best	Minimal
11	-	$1.78{\cdot}10^{-4}$	$1.11{\cdot}10^{-2}$	$7.54{\cdot}10^{-5}$ (1)	TG-FA1
12	-	$1.90{\cdot}10^{-5}$	$8.35{\cdot}10^{-3}$	$1.10{\cdot}10^{-3}$ (4)	OSA
13	-	$7.00{\cdot}10^{-4}$	$2.57{\cdot}10^{-2}$	$3.52{\cdot}10^{-5}$ (3)	TG-FA3
14	$1.51{\cdot}10^{-3}$	$6.7{\cdot}10^{-5}$	$1.07{\cdot}10^{-1}$	$6.01{\cdot}10^{-4}$ (1)	OSA
15	$1.2{\cdot}10^{-4}$	$1.39{\cdot}10^{-5}$	$3.28{\cdot}10^{-3}$	$6.41{\cdot}10^{-4}$ (1)	OSA

Notice that it was not possible to reconstruct the microtomographic samples using the PSA method for 3 of the 5 samples. This was due to the absence of connectivity in the final reconstruction. This result indicates that the correlation, DBI distribution, and chord size parameters were not sufficient to maintain 3D medium connectivity for samples with porosities between 9 % and 15 %. Comparing, also, the OSA* and OSA methods, it is noted that the former did not show good results. The reason for this will be discussed in the following section.

6.2.2. Considerations of Grain Size Distribution Correction

In this section, we present discussions and considerations about the methodology of correction of the distribution of grain size 2D to 3D.

It is noted in Figure 42 that the correlation curve measured in the reconstruction of sample 13 by the OSA* method was not reproduced adequately in the reconstruction performed. This result was similar for the other samples.

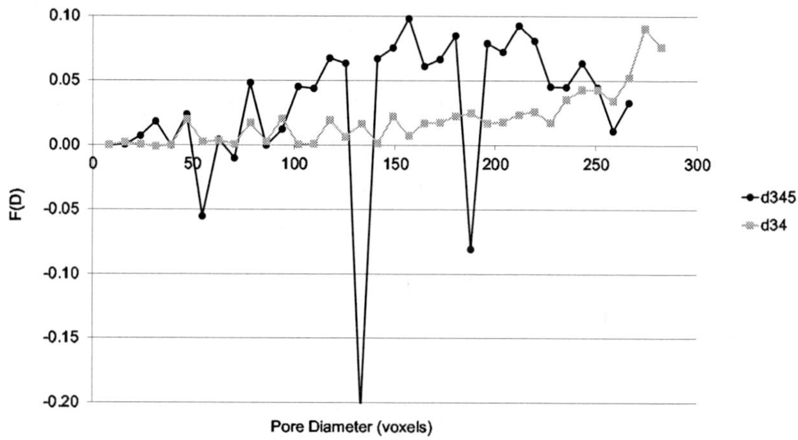

Figure 42. Comparison between the correlation functions of Sample 13 and a reconstruction using the method OSA*.

This curve showed a quadratic error of the order of $10^{-}3$, while the other methods have quadratic errors of the order of $10^{-}5$. The poor performance of the method would indicate a poor configuration of the Simulated Annealing parameters, causing a freeze to a local minimum. However, when the simulation was repeated altering the temperature parameters, the results were similar.

It was then observed the grain size distribution that was used as input to the reconstruction process compared to the grain size distribution measured in the microtomographic image. This comparison can be seen in Figure 43.

In this Figure, it can be noted that the measured 2D distribution has a larger grain size than the grains measured by the 3D distribution. This fact invalidates the corrections made, since the principle that all 3D grain sizes must be represented in a 2D distribution is no longer valid, and the consequence of this is that the corrected distribution has the same behavior as the 2D distribution and

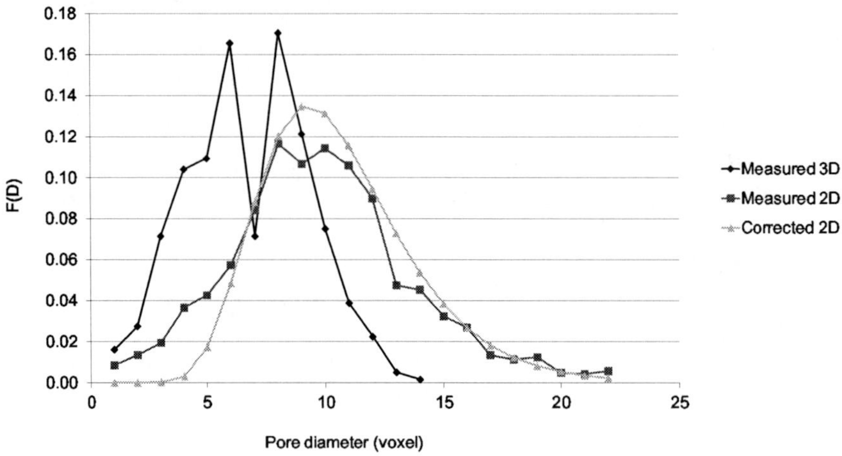

Figure 43. Comparison between the 2D and 3D grain size distributions measured in a microtomographic image through mathematical morphology and their respective corrected distribution.

generates a distribution with larger grains than they should actually be.

This distortion in the 2D grain size distribution occurs due to the mask used to perform the morphological opening operation (process used to measure the grain size distribution) to consider the grains as spherical, which does not occur in practice. In order to exemplify this effect, the geometry of Figure 44 was taken, where two spheres of rays 35 and 33 were centered at points (50,50,60) and (50,40,60), respectively, and this caused a partial overlap of both.

When measuring the grain size distribution (colored phase) through mathematical morphology, the data shown in Figure 45 was obtained. Note the above problem that the largest grain measured in the geometry section is larger than the 3D grain measured in the entire geometry. In addition, some values of the 3D distribution were observed to be negative. This effect is also a consequence of the measurement by mathematical morphology and should be disregarded.

These results made the use of grain size distribution correction unfeasible for reconstruction methods using Simulated Annealing. The following results will deal only with the OSA method, that is, without correction of the grain size distribution - using as input the 2D grain size distribution.

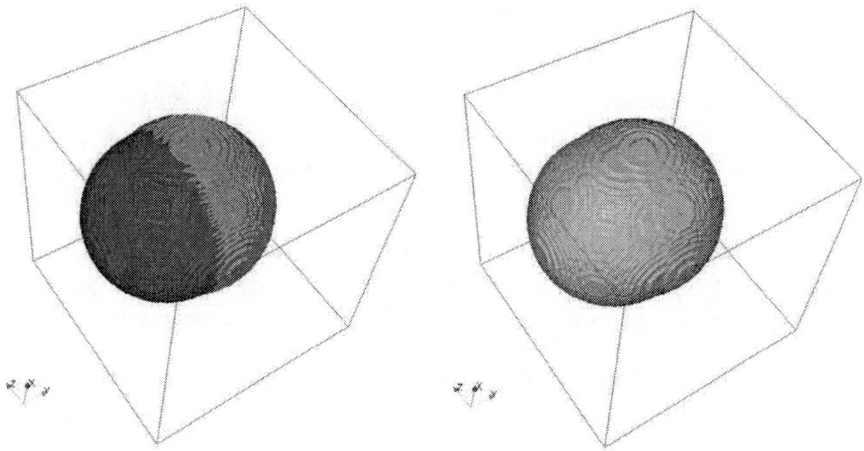

Figure 44. 3D geometry of two superimposed spheres of rays 35 and 33, centered on points (50,50,60) and (50,40,60), respectively.

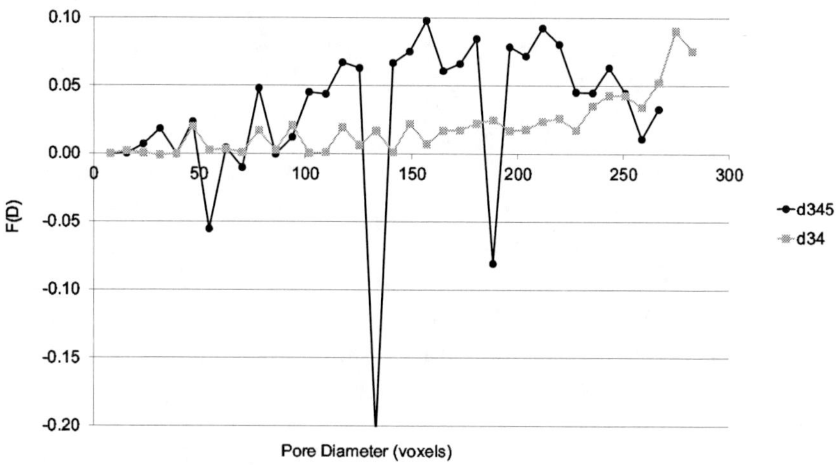

Figure 45. Grain Size Distributions (gray) and 3D (black) of Figure Geometry 44.

6.2.3. Pore Size Distribution

Another result to be analyzed is the 3D pore size distribution using the d_{3-4-5} metric. This measurement was performed on the 3D images, while the 2D measurements are made in the sections of 3D images. The Table 8 shows the error results for the analyzed samples. These results were measured throughout the porous space, not just the connected porosity. This methodology was chosen so that the TG method was not at a disadvantage in relation to the others in the samples of lower porosity, in which the method presented low connectivity.

Table 8. Comparison of the 3D pore size distribution between the reconstruction of 5 samples of porous media using the methods PSA, OSA, TG and SS

	Error regarding the d_{3-4-5} pore size distribution				
Sample	**PSA**	**OSA**	**TG-Best**	**SS**	**Minimal**
11	-	$1.66{\cdot}10^{-2}$	$4.84{\cdot}10^{-3}$ (1)	$2.21{\cdot}10^{-1}$	TG-1
12	-	$3.71{\cdot}10^{-2}$	$6.81{\cdot}10^{-2}$ (2)	$1.71{\cdot}10^{-1}$	OSA
13	-	$1.86{\cdot}10^{-2}$	$1.11{\cdot}10^{-1}$ (3)	$3.11{\cdot}10^{-2}$	OSA
14	$1.75{\cdot}10^{-1}$	$5.92{\cdot}10^{-2}$	$4.11{\cdot}10^{-2}$ (1)	$4.69{\cdot}10^{-2}$	TG-1
15	$3.30{\cdot}10^{-1}$	$2.37{\cdot}10^{-2}$	$3.24{\cdot}10^{-3}$ (1)	$5.70{\cdot}10^{-2}$	TG-1

The PSA method presented low performance again (only samples 14 and 15, were reconstructed with connected porosity), reaffirming that in addition to the connectivity, the adopted parameters were not enough to preserve the 3D pore size distribution. It is observed that the OSA method had results lower than the TG method, and both were higher than the SS, for the most part. Observing Figure 46, the pore size distribution of Sample 15 shows that the OSA method presented a curve indicating smaller pores than the other methods with the exception of ES. The same occurred with the sample 14, also of greater porosity.

On the other hand, samples 12 and 13, considered of low porosity, were better reconstructed by the OSA method, even when comparing the pore size distribution. One can see the pore size distribution measured in the microtomography of sample 13 and their respective reconstructions in Figure 47.

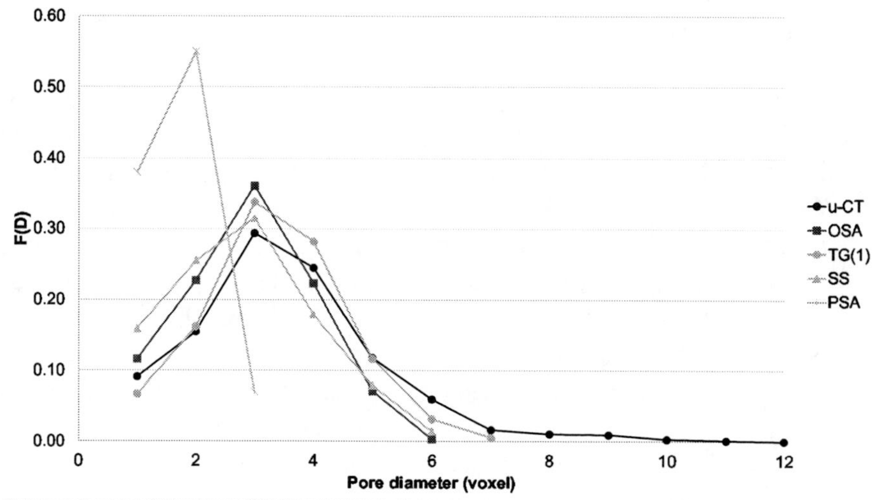

Figure 46. Comparison between the pore size distributions measured in the reconstructions using the methods TG(1), OSA, SS and the X-Ray microtomographic image for sample 15.

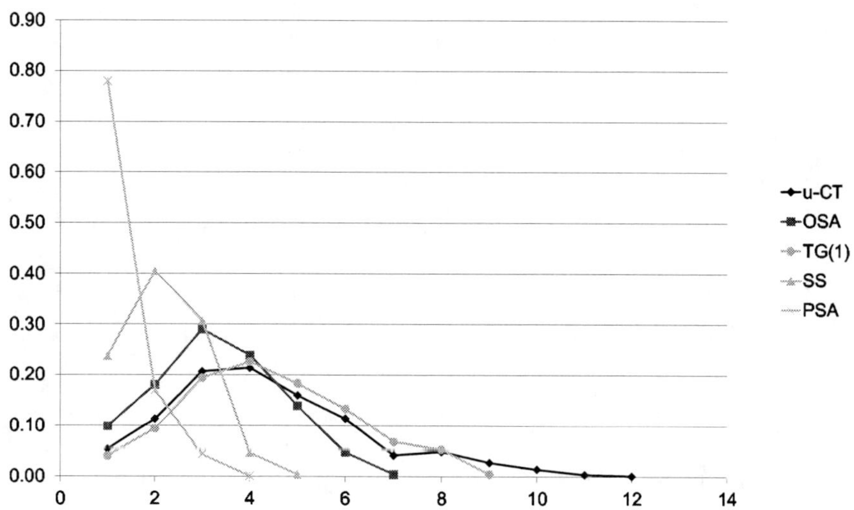

Figure 47. Comparison between measurement size distributions in reconstructions GT(1), OSA, SS and the X-Ray microtomographic image for sample 13.

Conclusion

It was studied a methodology to model particle size distributions. A SA algorithm was used to fit a 3D distribution with virtual cuts and minimizing a 2D function, comparing this latter with an input grain size distribution obtained experimentally.

It was tested 4 distribution functions, LogNormal, Fischer-Tippet, Shifted-Gompertz and LogLogistic. The function which had best performance was the Log-Logistic distribution, yet, in a simple test of flexibility, it was clear that the performance of the function fitting was not adequate.

A solution was proposed, which is to use a SA algorithm to freely fit a constant function. The preliminary tests showed that this algorithm was able to reproduce both a 3D Log-Normal function and a bimodal input function, indicating that it can be used to represent 3D particle size distributions determined from 2D particle size distributions.

It is also concluded that:

Given the complexity of the microstructures of the studied porous media, the correlation function by itself does not present enough information to rebuild a microstructure satisfactorily, especially for low porosity samples. These results confirm those presented by [Okabe and Blunt, 2004]. It is necessary to preserve higher order moments or other auxiliary parameters such as the linear string distribution and the fraction of percolating cells to obtain a reconstruction representing the original porous media.

The proposed grain size correction modeled the grain size distribution through an equation with adjustment of its variables by the technique of optimization of Simulated Annealing. The correction could not be used in this work due to problems in the 2D distribution through image analysis. As the mask used for the opening operation considers the spherical grain, which does not occur in practice, the grains measured in 2D were larger than those measured in 3D, which made the use of the proposed correction unfeasible.

Two geometric parameters were analyzed to complement the correlation: the linear chord distribution, with the objective of maintaining connectivity at short distances, and the DBI distribution, in order to preserve the pore size distribution of the microstructure; these parameters did not present satisfactory results in preserving the connectivity of the pore phase in the resulting reconstruction.

Two reconstruction methods were presented based on the Simulated Annealing optimization method: PSA, based on pixel motion, and OSA, which moves objects.

The PSA method was able to reconstruct synthetic structures with highly organized shapes, such as squares or cubes, through correlation parameters, linear chord distribution and DBI distribution and should be used for these type of structures, but it was not effective for the reconstruction of unorganized porous media, since the reconstruction did not preserve connectivity, and had poor performance in reproducing the pore size distribution, despite preserving the DBI distribution.

The OSA method presented improvements in relation to the other methods to reconstruct porous media, mainly in the parameter pore size distribution, in which the pores of larger diameter reproduced better than the other methods, and for the reconstruction with good connectivity of porosity samples less than 15%.

An extensive analysis of geometric parameters was performed to validate the reconstructions performed using both methods in relation to the Truncated Gaussian (TG) and Superposed Spheres (SS) reconstruction methods. The geometric parameters compared were the correlation function and pore size distribution.

It was observed that the set of three geometric parameters previously mentioned should be analyzed to verify which is the best reconstruction method to be used in order to adequately represent the microstructure of the porous media in which it is desired to determine the physical properties. The pore size distribution was an important parameter in the determination of these properties, as shown in the results presented, since the reconstructions that, even reproducing the correlation function, had too small pores, did not present good results in the determinations of properties physics. For the samples used in this work, the OSA method also proved to be the most appropriate, by combining the three parameters, especially for low porosity samples.

References

[Abercrombie, 1946] Abercrombie, M. (1946). Estimation of nuclear population from microtomic sections. *Anatomical Records*, 94(2).

[Abramson et al., 1999] Abramson, D., Dang, H., and Krisnamoorthy, M. (1999). Simulated annealing cooling schedules for the school timetabling problem. *Asia-Pacific Journal of Operational Research*, 16:1–22.

[Adler et al., 1990] Adler, P. M., Jacquin, C. G., and Quiblier, J. A. (1990). Flow in simulated porous media. *International Journal of Multiphase Flow*, 16(4):691–712.

[Appoloni et al., 2005] Appoloni, C. R., Rodrigues, C. R. O., and Fernandes, C. P. (2005). Porous microstructure characterization of a sandstone reservoir using high-resolution x-ray microtomography. In *2005 International Symposium of the Society of Core Analysts*.

[Arpat and Caers, 2007] Arpat, G. and Caers, J. (2007). Conditional simulation with patterns. *Mathematical Geology,*, 39(2).

[Baddeley, 2001] Baddeley, A. J. (2001). Is stereology 'unbiased'? *Trends in Neuroscience*, 24(7).

[Bain and Englehardt, 1992] Bain, L. and Englehardt, M. (1992). *Introduction to Probability and Mathematical Statistics*. Duxbury Press, Belmont, CA.

[Bartheld, 2002] Bartheld, C. V. (2002). Counting particles in tissue sections: choices of methods and importance of calibration to minimize biases. *Histological Histopathology*, 17(2).

[Bekri et al., 2003] Bekri, S., Howard, J., J. Muller, J., and Adler, P. (2003). Electrical resistivity index in multiphase flow through porous media. *Transport in Porous Media*, 51:41–65.

[Bekri et al., 2000] Bekri, S., Xu, K., Yousefian, F., Adler, P., Thovert, J.-F., Muller, J., Iden, K., Psyllos, A., Stubos, A., and Ioannidis, M. (2000). Pore geometry and transport properties in north sea chalk. *Journal of Petroleum Science and Engineering*, 25(3-4):107–134.

[Berryman and Blair, 1986] Berryman, J. G. and Blair, S. C. (1986). Use of digital image-analysis to estimate fluid permeability of porous material - application of two-point correlation function. *J. Appl. Phys.*, 60(6).

[Biswal et al., 1998] Biswal, B., Manwart, C., and Hilfer, R. (1998). Three dimensional local porosity analysis of porous media. *Physica A*, 266:307–311.

[Bostanabad et al., 2016] Bostanabad, R., Bui, A. T., Xie, W., Apley, D., and Chen, W. (2016). Stochastic microstructure characterization and reconstruction via supervised learning. *Acta Materialia*, 103:89–102.

[Cerny, 1985] Cerny, V. (1985). Thermodynamical approach to the traveling salesman problem: An efficient simulation algorithm. *Journal of Optimization Theory and Application*, 45:41–51.

[Chalkley, 1943] Chalkley, H. W. (1943). Methods for quantitative morphological analysis of tissue. *Journal of Natural Cancer Institute*, 4(1).

[Chalkley et al., 1949] Chalkley, H. W., Cornfield, J., and Park, H. C. (1949). A method for estimating volume-surface ratios. *Science*, 110.

[Chassery and Montanvert, 1991] Chassery, J. and Montanvert, A. (1991). *Geometrie Discrete en Analyse d' Images [Discrete Geometry in Image Analysis]*. Hermes.

[Coster, 1989] Coster, M.; Chermant, J. (1989). *Precis D'analyse D'images [Accurate Image Analysis]*. Presses du CNRS.

[Cule and Torquato, 1999] Cule, D. and Torquato, S. (1999). Generating random media from limited microstructural information via stochastic optimization. *Journal of Applied Physics*, 86(6):3428–3437.

[DeHoff and Rhines, 1968] DeHoff, R. T. and Rhines, F. N. (1968). *Quantitative Microscopy*. McGraw-Hill Book Company.

[Derossi et al., 2014] Derossi, A., Severini, C., and Pilli, T. D. (2014). Reconstruction of food microstructure via statistical correlation functions. the use of lineal-path distribution functions. *Journal of Food Engineering*, 142:9–16.

[dos Santos et al., 2002] dos Santos, L. O. E., Philippi, C., Fernandes, C. P., and de Gaspari, H. (2002). Reconstruo tridimensional de microestruturas porosas com o mtodo de esferas sobrepostas. In *Proceedings of the ENCIT 2002, Caxambu - MG, Brazil - Paper CIT02-0449*.

[Fernandes, 2002] Fernandes, C. (2002). *Monografia para o concurso de professor adjunto do departamento de engenharia mecnica da ufsc na rea de caracterizao de microestruturas com tcnicas de anlise de imagens. [Monograph for the assistant professor contest of the Department of Mechanical*

Engineering of the UFSC in the Microstructure Characterization Area with image analysis techniques] Florianpolis-SC.

[Garcia et al., 2007] Garcia, Y., Breen, A., Burugapalli, K., Dockery, P., and Pandit, A. (2007). Stereological methods to assess tissue response for tissue-engineered scaffolds. *Biomaterials*, 28(2).

[Gardella et al., 2003] Gardella, D., Hatton, W. J., Rind, H. B., Glenn, G. D., and Bartheld, C. S. V. (2003). Differential tissue shrinkage and compression in the z-axis: implications for optical disector counting in vibratome-, plastic- and cryosections. *Journal of Neuroscience Methods*, 124(1).

[Geman and Geman, 1984] Geman, S. and Geman, D. (1984). Stochastic relaxation, gibbs distributions, and the bayesian restoration of images. *IEEE Transactions on Pattern Analysis and Machine Intelligence*, 6:721–741.

[Gokhale, 1990] Gokhale, A. (1990). Unbiased estimation of curve length in 3d using vertical slices. *Journal of Microscopy*, 159.

[Gundersen, 1988] Gundersen, H. J. (1988). The nucleator. *Journal of Microscopy*, 151.

[Hajek, 1985] Hajek, B. (1985). A tutorial survey of theory and applications of simulated annealing. In *Proceedings of 24th IEEE Conference Decision and Control*, pages 755–760, Ft. Lauderdale, FL.

[Hedreen, 1999] Hedreen, J. C. (1999). Unbiased stereology? *Trends in Neuroscience*, 22(8).

[Hilfer, 2000] Hilfer, R. (2000). Local porosity theory and stochastic reconstruction for porous media. *Statistical Physics and Spatial Statistics*, 554:203–241.

[Howard et al., 1985] Howard, C. V., Reid, S., Baddeley, A., and Boyde, A. (1985). Unbiased estimation of particle density in the tandem scanning reflected light microscope. *Journal of Microscopy*, 138.

[Ioannidis and Chatzis, 2000] Ioannidis, M. A. and Chatzis, I. (2000). On the geometry and topology of 3d stochastic porous media. *Journal of Colloid and Interface Science*, 229:323–334.

[Jensen and Gundersen, 1993] Jensen, E. B. V. and Gundersen, H. J. G. (1993). The rotator. *Journal of Microscopy*, 170.

[Kirkpatrick et al., 1983] Kirkpatrick, S., Gelatt, C., and Vecchi, M. (1983). Optimization by simulated annealing. *Science*, 220:45–54.

[Laarhoven, 1988] Laarhoven, P. J. (1988). *Theoretical and Computational Aspects of Simulated Annealing*. Stichting Mathematisch Centrum, Amsterdam.

[Larsen and Nielsen, 1998] Larsen, J.O.; Gundersen, H. and Nielsen, J. (1998). Global spatial sampling with isotropic virtual planes: estimators of length density and total length in thick, arbitrarily orientated sections. *Journal of Microscopy*.

[Liang, 1997] Liang, Z. (1997). *Computer Generation and Application of 3-D Reconstructed Porous Structure: From 2-D Images to the Prediction of Permeability*. Tese de doutorado, Universidade Federal de Santa Catarina, Florianpolis-SC.

[Lundy and Mees, 1986] Lundy, M. and Mees, A. (1986). Convergence of an annealing algorithm. *Mathematical Programming*, 34:111–124.

[Mandarim-De-Lacerda, 2003] Mandarim-De-Lacerda, C. A. (2003). Stereological tools in biomedical research. In *Anais da Academia Brasileira de Cincias*, volume 75.

[Manwart et al., 2000] Manwart, C., Torquato, S., and Hilfer, R. (2000). Stochastic reconstruction of sandstones. *Physical Review E*, 62(1):893–899.

[Metropolis et al., 1953] Metropolis, N., Rosenbluth, A., Rosenbluth, M., Teller, A., and Teller, E. (1953). Equation of state calculations by fast computing machines. *Journal of Chemical Physics*, 21:1087–1092.

[Narayanan and Deans, 1988] Narayanan, K. and Deans, H. A. (1988). A flow model based on the structure of heterogeneous porous media. In *63rd Annual Technical Conference and Exibition of the Society of Petroleum Engineers*.

[Okabe and Blunt, 2004] Okabe, H. and Blunt, M. (2004). Prediction of permeability for porous media reconstructed using multiple-point statistics. *Physical Review E*, 70:1–10.

[Otten and Ginneken, 1989] Otten, R. and Ginneken, L. (1989). *The Annealing Algorithm*. Kluwer Academic Publishers.

[Ouenes and Bhagavan, 1994] Ouenes, A. and Bhagavan, S. (1994). Application of simulated annealing and other global optimization methods to reservoir description: Myths and realities. In *Proceedings of the SPE 69th Annual Technical Conference and Exhibition hold*, pages 547–561.

[Papoulis, 1991] Papoulis, A. (1991). *Probability, Random Variables, and Stochastic Processes, 3rd Edition*. Polytechic Institute of New York.

[Pazdniakou and Adler, 2013] Pazdniakou, A. and Adler, P. (2013). Dynamic permeability of porous media by the lattice boltzmann method. *Advances in Water Resources*, 62:292–302.

[Politis et al., 2008] Politis, M., Kikkinides, E., Kainourgiakis, M., and Stubos, A. (2008). A hybrid process-based and stochastic reconstruction method of porous media. *Microporous and Mesoporous Materials*, 110.

[Rintoul and Torquato, 1997] Rintoul, M. D. and Torquato, S. (1997). Reconstruction of the structure of dispersions. *Journal of Colloid and Interface Science*, 186:467–476.

[Romeo and Sangiovanni-Vincentelli, 1991] Romeo, F. and Sangiovanni-Vincentelli, A. (1991). A theoretical framework for simulated annealing. *Algorithmica*, 6:302–345.

[Serra, 1983] Serra, J. (1983). *Image Analysis and Mathematical Morphology*. Academic Press, Inc.

[Sheenan and Torquato, 2001] Sheenan, N. and Torquato, S. (2001). Generating microstructures with specified correlation functions. *Journal of Applied Physics*, 9(1):53–60.

[Sterio, 1984] Sterio, D. C. (1984). The unbiased estimation of number and sizes of arbitrary particles using the disector. *Journal of Microscopy*, 134.

[Stoicaa et al., 2005] Stoicaa, R. S., Gregorib, P., and Mateu, J. (2005). Simulated annealing and object point processes: Tools for analysis of spatial patterns. *Stochastic Processes and their Applications*, 115:1860–1882.

[Talukdar and Torsaeter, 2002] Talukdar, M. and Torsaeter, O. (2002). Reconstruction of chalk pore networks from 2d backscatter electron micrographs using a simulated annealing technique. *Journal of Petroleum Science and Engineering*, 33:265 – 282.

[Torquato, 1992] Torquato, S.; Lu, B. (1992). Lineal-path funtion for random heterogeneous materials:. *Physical Review A*, 45(2).

[Torquato, 2002] Torquato, S. (2002). *Random Heterogeneous Materials: Microstucture and Macroscopic Properties*. Springer-Verlag.

[Torquato and Kim, 1999] Torquato, S. and Kim, I. (1999). Effective conductivity, dielectric constant, and diffusion coefficient of digitized composite media via first-passage-time equations. *Journal of Applied Physics*, 85:1560.

[Torquato and Lu, 1993] Torquato, S. and Lu, B. (1993). Chord-length distribution for a two-phase random media. *Physical Review E*, 4(4):2950–2953.

[Varanelli, 1996] Varanelli, J. (1996). *On the Acceleration of Simulated Annealing*. PhD thesis, University of Virginia.

[Venkatarangan, 2000] Venkatarangan, A. B. (2000). *Geometric and Statistical Analysis of Porous Media*. PhD thesis, State University of New York.

[Weibel and Gomez, 1962] Weibel, E. R. and Gomez, D. M. (1962). A principle for counting tissue structures on random sections. *Journal of Applied Physiology*, 17.

[Yeong and Torquato, 1998] Yeong, C. L. Y. and Torquato, S. (1998). Reconstructing random media. *Physical Review E*, 57(1).

In: Simulated Annealing
Editors: Alex Scollen et al.
ISBN: 978-1-53613-674-6

Chapter 6

A Case Study of Energy Recovery Systems Location in Irrigation Networks Defined by Simulated Annealing Techniques

***Modesto Pérez-Sánchez*[1], *Francisco-Javier Sánchez-Romero*[2] *and P. Amparo López-Jiménez*[1]**

[1]Hydraulic and Enviromental Engineering Departament, Universitat Politècnica de València, Valencia, Spain
[2]Rural and Agrifood Engineering Departament, Universitat Politècnica de València, Valencia, Spain

Abstract

Increase of sustainability in any system must be considered by all managers of different infraestructures as a primary objective. This concept consideration inside of water systems is also necessary. Therefore, water managers have to look for news strategies in order to improve the hydraulic and energy efficiency of their water networks. In this sense, changing pressure reduction valves by pumps working as turbines has been studied by researches since 1990s. However, the study of this recovery machines also needs strategies to help water managers to allocate them, maximizing the recovered energy. In this research, a maximization methodology to recover energy, considering installation feasibility, was proposed to allocate recovery manchines within water networks, by using simulated annealing techniques Combination of Epanet (hydraulic code) and simulated annealing in the proposed methodology is a powerful water management tool towards the search of the sustainability in irrigation networks. Particularly, by usingthis methodology 186 MWh/year could be thoretically recovered, reaching payback below 9 years. This energy generation could contribuite to develop more sustainable water systems, taking adavantageof optimization techniques.

Keywords: simulated annealing, PATs, sustainable water systems, smart irrigation networks

INTRODUCTION

The need to increase the sustainability in different processes is causing management change of the infrastructures. This new way of thinking is also involving water systems, big energy consumers. As an example, the consumed energy in water distribution represented 7% of the worldwide used energy (Pérez-Sánchez et al., 2016). This change of management has been focused in different strategies for last years. Methodologies varied depending on water systems. In pumped systems, the increase of the energy is related to supply uniform area with similar level, reducing the manometric head. In gravity systems, the increase of the efficiency is obtained by pressure reduction. This reduction enables to increase hydraulic performance of the network as well as energy efficiency since each unit distributed volume consumes energy (Castro, 2006).

Pressure reduction has traditionally been developed by using valves for last decades, but Ramos and Borga (1999) proposed a sustainable solution in which, the use of pumps working as turbines (PATs) enabled to reduce the pressure and to generate energy, taking advantage of the reduction of the manometric head. However, the improvement of energy efficiency in water systems is perfect if the energy recovery is maximum. To get the best result, optimization techniques must be implemented. In this line, Perez-Sánchez et al., (2018) developed a short literature review about different optimization procedures with different complexity. All strategies were focused on maximizing recovery energy as well as minimizing excess of pressure (Giugni et al., 2014). Araujo et al., (2006) used stochastic methods for using PRVs in water systems. Prats et al., (2011) minimized the use of pumped system. However, the impossibility to get optimal solutions caused the use of metaheuristic algorithms such as simulated annealing (Samora et al., 2016; Pérez-Sánchez et al., 2018), PSO (particle swarm optimization) (Kaviani et al., 2009) or tabu search (Cunha and Ribero, (2004).

Optimization developed by an analytical strategy, was carried out in two phases. First one was focused on integral balance of energy in a control volume (White, 2008). Energy terms were defined, distinguishing between recoverable and non-recoverable energy, depending on the required energy for irrigation. Second phase was proposed to develop an optimization strategy through a heuristic algorithm, based on analogy with the physics process of annealing of metals and inspired in Monte-Carlo's (Kirkpatrick et al., 1983). However, the recovered energy by a recovery system depends on efficiency of the implemented hydraulic machine. Furthermore, the regulation is crucial to maximize recovered energy (Pérez-Sánchez et al., 2017) as flow and head are variable (Pérez-Sánchez et al., 2016). Therefore, variation of head and efficiency curves of the PATs (i.e., when flow changes) is crucial to carry out an optimization of recovered energy in real water distribution systems.

The present research describes a real energy recovery study using an optimization procedure developed by Pérez-Sánchez et al., (2017). This analysis considered many factors, such as: (1) the farmers' demands, maximum days between irrigations, weekly trend of irrigation, and irrigation duration; (2) location of the PATs in order to maximize recovered energy; (3) selected machine, depending on specific speed; (4) rotational speed of the machine, causing variation of efficiency and recovered head for each flow;(5) feasibility of the machines installation in real water distribution networks.

MATERIALS AND METHODS

Recovered energy depends on several constrains, related to consumption points. These main variables are: input of energy in the water system; demand patterns; pressure and flow requirements; and finally, size of pipelines of systems, since losses joint to consumption flow determine friction energy consumed on distribution. Pérez-Sánchez et al., (2018) defined recovered energy by equation (1):

$$E_{R_i}(kWh) = \sum_{k=1}^{k=p} \eta\gamma Q_{ik}\left(p_{ik} - \max\left(p_{min_i}; p_{minI_i}\right)\right)\Delta t = \sum_{k=1}^{k=p} \eta\gamma Q_{ik} H_{ik} \Delta t \quad (1)$$

where E_{R_i} is recovered energy in line 'i' (kWh); 'R', means that the energy recovery value is obtained with a defined machine; *i* is the line of the system where turbine is installed, which varies between 1 and '*n*'; *k* is time interval analyzed, which varies between 1 and '*p*'; *p* is the number of analyzed intervals; η is machine efficiency; γ is specific weight of fluid (N/m^3); Q_{ik} is circulating flow by a line i(irrigation point, hydrant or line), in interval k (m^3/s); p_{ik} is service pressure in a line i, in interval k (m w.c); p_{min_i} is minimum pressure of line or hydrant to ensure the minimum pressure in the most disadvantageous consumption node (m w.c.); p_{minI_i} is minimum required pressure of an irrigation point to ensure evenly irrigation water (m w.c.); H_{ik} is head value in the line i, in interval k (m w.c); Δt is time interval (s). If the total time analyzed is a year, units for E_{R_i} are kWh/year.

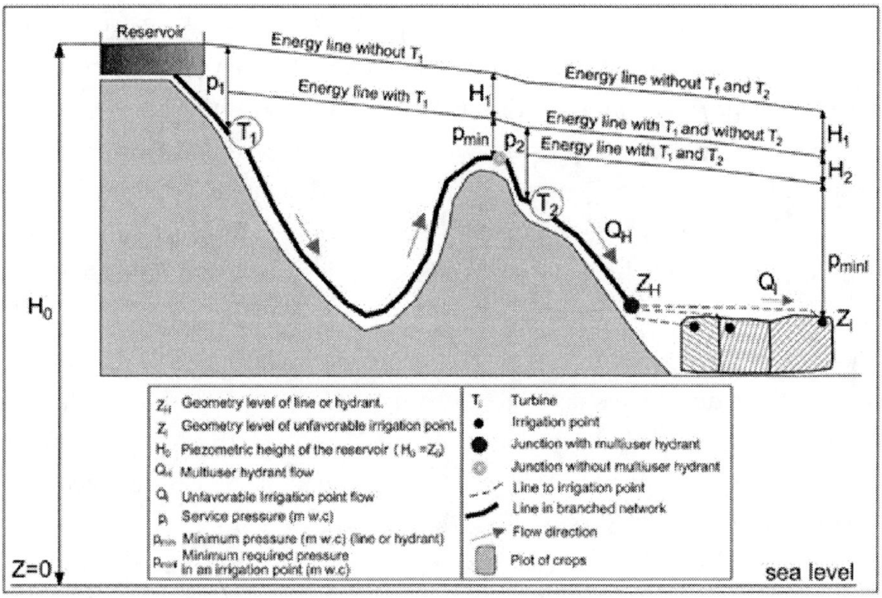

Figure 1. Network with 2 turbines that are installed on serie.

If different PATs are installed on series (Figure 1), the recovered energy in a line *i* is defined by equation (2):

$$E_{Ri}^{S} = \sum_{k=1}^{k=p} \eta\gamma Q_{ik}\left(H_{ik} - \sum_{i=1}^{i=m}\left(\mathrm{H}_{ik}^{S}\right)\right)\Delta t \qquad (2)$$

where *m* is the number of installed turbines upstream, varying between 0 and '*n-1*'; *n* is the total number of installed machines in the water network; E_{Ri}^{S} is recovered energy by the turbine that is installed in the line '*i*' with other turbines installed in series (kWh); H_i^S is recovered head by other upstream installed turbines in series (m w.c.).

If a network has a total number of lines equal to *i* and *n* turbines, the total energy recovered (E_{TR}) is calculated by equation (3):

$$E_{TR} = \sum_{j=1}^{j=n}([A_j] \cdot [B] \cdot [C_j]) \qquad (3)$$

where $[A_j]$ is the matrix (1 x i) made up zero or one. This matrix defines where turbine j is installed. If the value is one, turbine j is installed in this line, otherwise, the line doesn't have any installed turbine. In this matrix only appears one value equal to one. $[B]$ is the matrix (i x i) which defines the value of recovered energy in the line, when another turbine exists upstream This matrix is calculated by equation (2) applied in all possible combinations of two lines on series. This matrix is constant in the network. $[C_j]$ is the matrix (i x 1) and its value is zero or one. The matrix defines the lines where another turbine is placed upstream the turbine j. If the value es one, another turbine exists in this position. If no more turbine is located upstream, the value in position j is one. These matrixes are generated according to the topology of the network. Equation (3) defines recovered energy for any configuration of n turbines in the network.

EPANET toolkit is used to determine pressure and flow over time for each node and line in order to get recoverable energy. EPANET is public domain software (Rossman, 2000).

OPTIMIZATION METHODOLOGY FOR MAXIMIZING ENERGY

Simulated Annealing (SA) is a combinatorial optimization method. The algorithm is inspired in Monte-Carlo's method (Kirkpatrick, Gelatt, & Vecchi, 1983) based on Metropolis's idea (Metropolis, Rosenbluth, Rosenbluth, Teller, & Teller, 1953).This method is a powerful tool to solve problems in which a high number of combinations is possible. The requirements to apply simulated annealing procedure in an engineering problem are: (1) configuration of decision variables, (2) definition of objective functions, (3) a procedure for generating rearrangements of the system, (4) control parameter and an annealing schedule, and (5) a criterion for termination (Dougherty & Marryott, 1991).

Configuration of Decision Variables

Decision variables consist to select the lines of branched network where recovery machines must be installed. The algorithm searches these locations in order to maximize recovered energy, finding an optimal configuration of lines. For each moment of the simulation, the method also confirms pressure and flow conditions that are verified for each consumption point. This rule must always be considered since energy recovery is not the main goal of water distribution systems (San Bruno, Choulot, & Denis, 2010).

Definition of Objective Functions

Two different functions are studied to analyze maximization of recovered energy in the network. The first function (Ψ_1) is recovered energy and it is defined by equation (4):

$$\Psi_1 = E_{TR} \tag{4}$$

Installation of a high number of recovery machines implies a greater recovered energy. However, the maximum number of installed turbines in a network should be established considering economic conditions that define the feasibility of the installation. This second function uses an economic criterion to define feasibility of the recovery system. (Castro, 2006; Zema, Nicotra, Tamburino, & Zimbone, 2016). The second function is defined by equation (5):

$$\Psi_2 = \frac{E_{TR}}{PSR} \tag{5}$$

where PSR is period simple return. This indicator is defined by equation (6):

$$PSR = \frac{IC}{I-C} \tag{6}$$

where IC is investiment cost (€), including the civil work, electric and electronic equipment. IC is defined by equation (7):

$$IC = C_i \cdot P \tag{7}$$

where C_i is estimated in 545 €/Kw (Castro, 2006); and P is the power of installed turbine. This power corresponds to the maximum power (P_{max}), and is defined by equation (8):

$$P = \eta\gamma Q_{ik} H_{ik} \tag{8}$$

where Q_{ik} and H_{ik} is circulating flow and head in line i and interval k, respectively when P_{max} is obtained. When E_{TR} and P are calculated for η is considered the value 0.55 according to (Castro, 2006).

I is the annual income (€/year) defined by equation (9):

$$I = P_E \cdot E_{TR} \tag{9}$$

where P_E is the selling energy price (€/kWh), estimating an average value of 0.0842 €/kWh; and C is the annual operating cost (€/year) defined by equation (10):

$$C = C_0 \cdot E_{TR} \tag{10}$$

where C_0 is the unit operation cost (€/kWh), estimating this parameter as 0.0145 €/kWh.

With these two objective functions, energy is considered by Ψ_1 and economical aspects (investment and economic viability) are considered by PSR in Ψ_2.

PROCEDURE TO APPLY SIMULATED ANNEALING METHOD

The maximization process is shown in Figure 2. A list of the lines that are ordered in descendent value of the objective function is developed, using the results in the hydraulic simulation according to the selected objective function. The values of the control parameters are assigned to characterize the simulated annealing process, defining the number of iterations. These parameters are: initial temperature (T_i), final temperature (T_f), cooling rate (α) and number of transitions (L_0). These parameters are fixed according to a sensitivity analysis previously performed. In this process each temperature transition (T_t) is calculated by geometric relation according to equation (11):

$$T_t = \alpha_c \cdot T_{t-1} \tag{11}$$

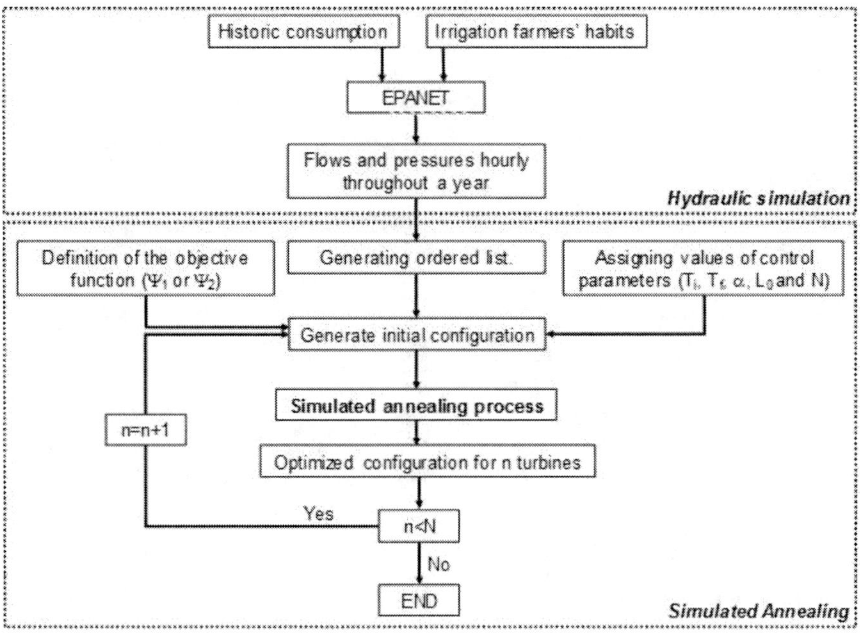

Figure 2. Flowchart for networks optimization in the energy recovery process with *n* installed turbines.

A maximum number of turbines (N) is necessary to be defined. The range for this value varies between 2 and the maximum number of lines in the water network.

The initial combination (X), $X = (x_1, x_2, \ldots.., x_n)$ is established, in which n is the number of turbines and x_i is the line where each turbine is located. Initially, the selected lines are the first two lines. When the initial configuration is created, the optimization process starts. In this process, each new combination is generated according to the general process of the simulated annealing method. For each iteration, a line of the combination is randomly deleted, and another line is randomly introduced. The probability introduced in the new combination is proportional to the value of the objective function.

When the process is finished and n is below N, the process starts with $n + 1$ turbines. Initial configuration for $n + 1$ turbines is created with the previous optimized configuration for n turbines and a new line is selected using the ordered list. The criterion to choose the new line is also proportional to the value of the objective function.

CASE STUDY

A real case study corresponding to a real irrigation network is presented, to apply the described methodology and determine the optimal number of turbines that could be installed.

The modelled network was located in Alicante (Spain), with an irrigated area of 289 ha and the main irrigated crop was citrus and loquats (Figure 3). Irrigation system was designed as an on-demand branched network (users always get the water resource with enough quality and quantity). The level difference between higher and lower irrigation point was 158 m and the reservoir ensured the minimum pressure that was 30 m w.c. in all consumption points. Diameter of pipelines varied between 54.4 and 700 mm and they were of different materials (ductile iron and polyvinyl-chloride). Total number of hydrants was 137, with 904 irrigation points. These consumption points were connected with hydrants by polyethylene pipelines.

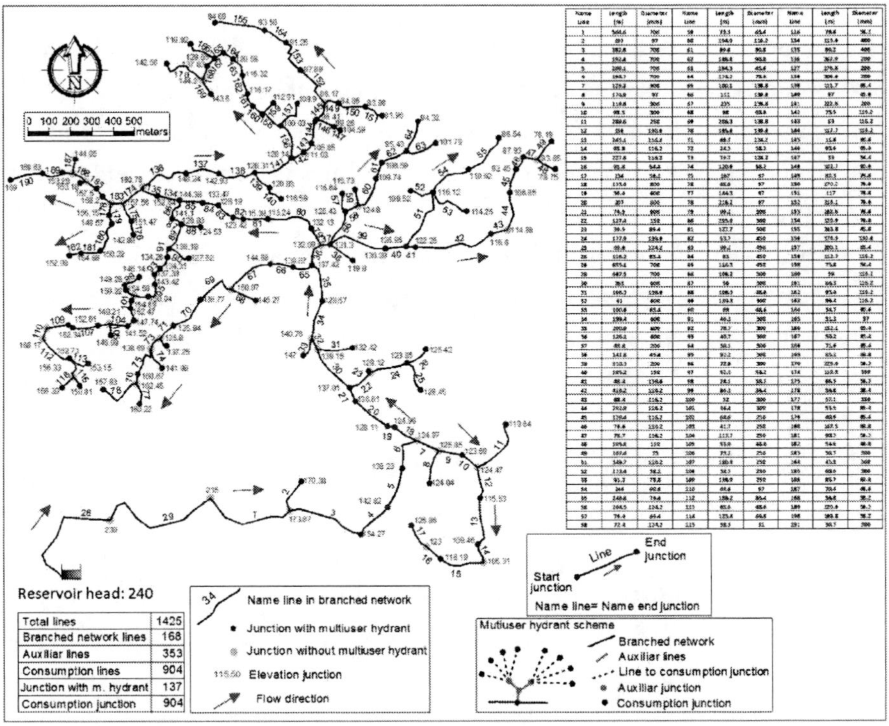

Figure 3. Case study.

If the proposed methodology is considered, the maximum number of turbines to install could be 168 in this network. Therefore, there are many combinations that are possible for each number of turbines installed. Nevertheless there is only one combination that maximizes objective function. In order to obtain this best solution, three stages were developed:

First step: Before allocating recovery machines, the flow were estimated for any line of the network. To determine the flow, the proposed methodology by Pérez-Sánchez (2016) was applied. This methodology calculates demand patterns according to farmer's habit (Pérez-Sánchez et al., 2016). In this case, the total number of developed patterns were 904 (i.e., one pattern for each irrigation point), which were hourly discretized over time.

The objective function was calculated for each line of the water network in the ordered lists, and Ψ_1 and Ψ_2 were generated. Values for Ψ_2 were developed considering the values of the recovered energy and the PSR.

Second step: To analyse the robustness of this proposed methodology, best optimized result was compared with the higher value considering all possible combinations (i.e., the position of turbines was considered in any line). For each objective function, the best 22 lines were chosen, and the best optimized solution was defined from 2 to 10 turbines. This step enabled to fix the different control parameters by a sensitivity analysis.

Third step: For each objective function, optimized solution in all considered cases were obtained from 2 to 168 lines.

RESULTS AND DISCUSSION

Table 1 shows the obtained values of the theoretical recovered energy. This values were sorted in descendent order. Line 4 had the higher recovered energy, with 131.76 MWh/year. For E_{TR}/PSR, the high value (line 20) was 25.14 MWh/year^2 and the minimum value was 3.74 year (line 32) when PSR was considered. The used control parameters in both objective functions for this case were: $T_i = 100$, $T_f = 0.01$, $\alpha = 0.9$ and t $L_0 = 10$. 792 iterations were performed for each optimization process considering the previous parameters.

Table 2 and Table 3 show the results for each objective function. When optimization process was applied considering the best 22 lines, the best solution in all considered cases from two to ten turbines was obtained for Ψ_1. When the Ψ_2 was calculated, results were also suitable. In this case, the maximum difference using simulated annealing was -0.31% compared to energy value when all combinations were developed. All possible combinations were determined by 1744413 configurations. The proposed optimization procedure determined 7128 (792 times 9) configurations to estimate de optimized solution. Both tables show the optimization process that got best solutions when they were compared with the obtained values.

Table 1. Ordered list considering the theoretical recovered energy lines (E_{TR})

Pipe	E_{TR}	PSR	$\frac{E_{TR}}{PSR}$	Pipe	E_{TR}	PSR	$\frac{E_{TR}}{PSR}$	Pipe	E_{TR}	PSR	$\frac{E_{TR}}{PSR}$	Pipe	E_{TR}	PSR	$\frac{E_{TR}}{PSR}$
4	131.76	5.33	24.72	42	12.81	6.23	2.06	15	5.72	12.19	0.47	157	1.79	11.86	0.15
5	131.16	5.34	24.56	65	12.03	5.97	2.02	48	5.68	12.27	0.46	57	1.79	14.79	0.12
6	130.12	5.34	24.37	144	12.03	6.53	1.84	103	5.48	5.69	0.96	49	1.75	16.83	0.10
3	126.47	5.61	22.54	44	12.02	6.26	1.92	164	5.36	6.40	0.84	139	1.72	13.56	0.13
1	120.07	5.94	20.21	43	11.67	6.45	1.81	71	5.28	7.29	0.72	2	1.71	18.58	0.09
13	106.88	4.32	24.74	158	11.61	4.80	2.42	104	5.18	5.91	0.88	146	1.68	12.43	0.14
19	104.91	4.24	24.74	45	11.49	6.74	1.70	149	5.05	8.71	0.58	166	1.64	11.61	0.14
20	101.63	4.00	25.41	12	11.44	9.50	1.20	165	4.97	6.72	0.74	59	1.63	12.62	0.13
21	100.57	4.00	25.14	24	11.28	11.04	1.02	106	4.67	5.91	0.79	188	1.56	5.97	0.26
30	91.18	3.97	22.97	56	11.20	6.82	1.64	54	4.57	10.08	0.45	116	1.54	6.80	0.23
32	87.17	3.74	23.31	145	11.08	6.82	1.62	152	4.51	7.82	0.58	88	1.51	9.30	0.16
34	85.51	3.81	22.44	89	10.99	5.42	2.03	73	4.43	8.12	0.55	97	1.47	8.38	0.18
35	82.69	3.89	21.26	66	10.94	5.92	1.85	150	4.33	9.60	0.45	147	1.37	14.13	0.10
35	71.76	4.12	17.42	91	9.99	5.62	1.78	25	4.20	17.35	0.24	159	1.33	12.59	0.11
79	47.39	3.97	11.94	160	9.80	4.65	2.11	183	3.87	5.88	0.66	181	1.31	7.85	0.17
80	45.86	3.87	11.85	92	9.63	5.43	1.77	64	3.72	11.12	0.33	189	1.31	6.34	0.21
81	45.67	3.89	11.74	58	9.29	7.13	1.30	75	3.72	8.66	0.43	169	1.29	15.37	0.08
82	45.28	3.90	11.61	93	9.19	5.65	1.63	153	3.68	8.10	0.45	78	1.28	13.83	0.09
83	44.91	3.92	11.46	51	9.16	8.12	1.13	55	3.61	12.68	0.28	68	1.14	11.14	0.10
84	44.72	3.89	11.50	67	9.16	5.96	1.54	178	3.54	6.55	0.54	16	1.14	9.72	0.12
85	44.05	3.91	11.27	148	9.09	6.72	1.35	107	3.47	5.60	0.62	17	1.14	9.72	0.12
7	38.41	13.32	2.88	94	9.01	5.59	1.61	8	3.29	13.59	0.24	72	1.09	15.08	0.07
29	36.99	5.87	6.30	161	8.95	4.83	1.85	184	3.28	6.94	0.47	63	1.06	11.91	0.09

Table 1. (Continued)

Pipe	E_{TR}	PSR	$\frac{E_{TR}}{PSR}$	Pipe	E_{TR}	PSR	$\frac{E_{TR}}{PSR}$	Pipe	E_{TR}	PSR	$\frac{E_{TR}}{PSR}$	Pipe	E_{TR}	PSR	$\frac{E_{TR}}{PSR}$
37	36.70	4.70	7.81	46	8.91	8.68	1.03	108	3.11	5.91	0.53	182	1.06	8.97	0.12
9	35.44	14.43	2.46	174	8.56	4.48	1.91	179	3.10	7.42	0.42	155	0.92	13.79	0.07
10	33.48	14.78	2.27	95	8.41	5.38	1.56	151	3.06	13.06	0.23	98	0.92	11.64	0.08
134	31.58	4.27	7.40	60	8.36	6.98	1.20	167	3.01	8.07	0.37	74	0.87	16.12	0.05
135	31.28	4.21	7.43	69	8.15	6.55	1.24	112	2.94	6.08	0.48	77	0.85	16.61	0.05
136	29.15	4.36	6.69	96	8.09	5.59	1.45	110	2.85	6.26	0.46	170	0.80	12.85	0.06
137	28.96	4.32	6.70	70	7.85	6.59	1.19	109	2.85	6.26	0.46	187	0.78	15.25	0.05
138	27.56	4.25	6.48	61	7.62	7.59	1.00	154	2.84	9.55	0.30	140	0.72	15.21	0.05
39	25.27	5.56	4.54	162	7.46	5.09	1.47	26	2.61	11.87	0.22	190	0.69	8.96	0.08
40	24.71	5.69	4.34	13	7.38	10.01	0.74	53	2.58	14.38	0.18	105	0.69	14.19	0.05
141	24.21	4.47	5.42	47	7.22	10.92	0.66	185	2.55	7.06	0.36	113	0.68	10.52	0.06
11	24.13	18.74	1.29	177	7.16	4.65	1.54	23	2.53	12.54	0.20	90	0.67	15.02	0.04
41	16.40	6.90	2.38	100	6.91	6.08	1.14	186	2.12	6.66	0.32	52	0.67	11.15	0.06
142	14.17	5.85	2.42	62	6.73	8.06	0.83	33	2.08	12.61	0.16	38	0.65	12.00	0.05
86	13.56	5.35	2.53	101	6.61	5.96	1.11	180	2.07	7.04	0.29	191	0.49	14.46	0.03
143	13.49	5.82	2.32	31	6.33	13.97	0.45	76	2.07	11.26	0.18	99	0.44	12.05	0.04
22	13.26	10.29	1.29	163	6.11	5.82	1.05	168	2.06	11.14	0.18	176	0.41	16.08	0.03
156	13.26	4.27	3.11	102	5.81	5.46	1.06	114	1.83	6.45	0.28	115	0.32	12.85	0.02
87	13.01	5.32	2.45	14	5.72	12.19	0.47	175	1.79	7.61	0.24	28	0.00	0.00	0.00

E_{TR} (MWh/year) PSR (year) E_{TR} /PSR (MWh/year^2)

Table 2. Comparison of the results for objective function Ψ_1

Objective function $\Psi_1 = E_{TR}\ (\frac{MWh}{year})$								
Number of turbines	All combinations with the best 22 lines		Simulated Annealing solution with the best 22 lines ($N = 10$)			Simulated Annealing solution with 168 lines ($N = 168$)		
	E_{TR}	Number of configurations	E_{TR}	Difference between SA and the best of all possible combinations (%)	Number of configurations to obtain optimized solution	E_{TR}	Difference between SA and the best of all possible combinations (%)	Number of configurations to obtain optimized solution
2	146.93	231	146.93	0.00	10	146.93	0.00	250
3	150.43	1540	150.43	0.00	24	159.51	+6.04	212
4	151.17	7315	151.17	0.00	23	166.95	+10.44	49
5	151.28	26334	151.28	0.00	48	171.70	+13.50	336
6	151.28	74613	151.28	0.00	146	175.74	+16.17	288
7	151.28	170544	151.28	0.00	432	179.88	+18.91	220
8	151.28	319770	151.28	0.00	751	182.37	+20.55	190
9	151.28	497420	151.28	0.00	776	183.84	+21.52	62
10	151.28	646646	151.28	0.00	790	185.74	+22.78	135

Table 3. Comparison of the results for objective function Ψ_2

Objective function $\Psi_2 = E_{TR}/PSR$ $(\frac{MWh}{year^2})$								
Number of turbines	All combinations with the best 22 lines		Simulated Annealing solution with the best 22 lines ($N = 10$)			Simulated Annealing solution with 168 lines ($N = 168$)		
	$\frac{E_{TR}}{PSR}$	Number of configurations	$\frac{E_{TR}}{PSR}$	Difference between SA and the best of all possible combinations (%)	Number of configurations to obtain optimized solution	$\frac{E_{TR}}{PSR}$	Difference between SA and the best of all possible combinations (%)	Number of configurations to obtain optimized solution
2	26.06	231	26.06	0.00	155	26.06	0.00	203
3	26.06	1540	26.06	0.00	2	26.89	+3.18	35
4	26.06	7315	26.02	-0.15	26	27.18	+4.30	193
5	26.02	26334	26.01	-0.04	689	27.56	+5.92	478
6	26.02	74613	25.97	-0.19	145	27.58	+6.00	110
7	26.01	170544	26.00	-0.04	738	27.56	+5.96	101
8	25.97	319770	25.97	0.00	789	27.56	+6.12	314
9	25.97	497420	25.92	-0.19	95	27.51	+5.93	562
10	25.96	646646	25.88	-0.31	411	27.48	+5.86	636

Table 4. Combinations for $\Psi_1 = E_{TR}$ maximization with N =168 lines

Number of turbines	Lines										Recovered Energy (MWh/year)	PSR (year)	$\frac{E_{TR}}{PSR}$ (MWh/year^2)
2	4	7									146.93	6.81	21.58
3	4	7	37								159.51	6.89	23.15
4	4	7	37	137							166.95	7.02	23.78
5	4	7	22	37	137						171.70	7.20	23.85
6	5	7	22	37	137	142					175.74	7.24	24.27
7	4	7	22	37	45	137	142				179.88	7.26	24.78
8	4	7	11	22	37	45	137	142			182.37	8.18	22.29
9	2	4	7	11	24	37	45	137	142		183.84	8.29	22.18
10	2	4	7	11	24	37	45	61	137	142	185.74	8.33	22.30

Table 5. Combinations for $\Psi_2 = E_{TR}/PSR$ maximization with N =168 lines

Number of turbines	Lines										Recovered Energy (MWh/year)	PSR (year)	$\frac{E_{TR}}{PSR}$ (MWh/year^2)
2	4	37									144.34	5.54	26.06
3	4	37	144								150.56	5.60	26.89
4	4	37	45	145							153.66	5.65	27.18
5	4	37	45	144	156						157.52	5.72	27.56
6	4	16	37	45	144	156					157.97	5.73	27.58
7	4	37	45	145	156	162	164				157.08	5.70	27.56
8	4	5	37	45	62	144	156	162			159.39	5.78	27.56
9	4	30	32	37	44	145	156	162	165		159.84	5.79	27.59
10	4	17	37	44	47	61	144	156	163	167	160.67	5.85	27.48

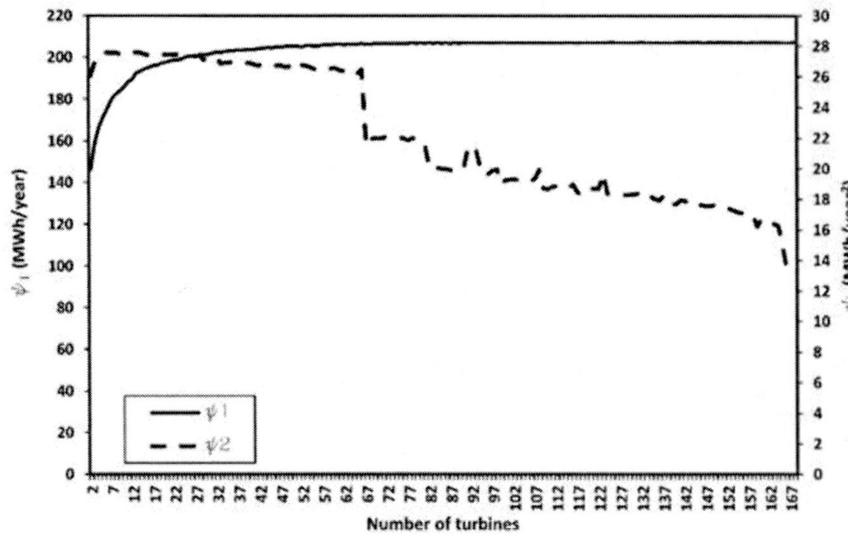

Figure 4. Results for all 168 lines for both objective functions.

When the optimization process was applied considering all lines (N =168 lines), the final values were improved. Increment of recovered energy oscillated between +6.04% (when three turbines were simulated) and +22.78% (when ten turbines were simulated). For Ψ_2, the maximum improvement was +6.12 (with eight turbines) and the minimum was +3.18% (with three turbines).

Table 4 and Table 5 show the best solution for each objective function. The maximum recovered energy and PSR values were calculated for different number of installed PATs, where the maximum number of considered machines was ten. For each combination, the selected lines were indicated in each table. Figure 4 shows obtained values for all lines (168) depending on each objective function. When Ψ_1 was analysed, the maximum energy was 207 MWh/year when 124 machines were installed. This solution was theoretical since the PSR was too bad particularly, the PSR value is 14.55 years when the PATs were installed. For Ψ_2, the maximum value was 27.59 MWh/year^2. This value was reached when nine turbines were installed. If the number of machines increased, the objective function decreased. Therefore, the consideration of PSR inside of the objective function was crucial to fix the number of machines, as well as to know the place where the recovery machines must be installed.

Figure 4 shows different behaviors for each objective function. If Ψ_1 is analyzed, it always grew, when 2 turbines the recovered energy was 146.93 MWh/year and the maximum value was 207 MWh/year. If the other objective function (Ψ_2) is considered, different areas were obtained. Firstly, from 2 to 9 turbines, Ψ_2 grew until the maximum value (27.59 MWh/year^2). In a second zone, from 9 to 67 turbines, Ψ_2 decreased softly. In the third zone, from 67 to 168 turbines, Ψ_2 quickly decreased to a minimum value of 13.54 MWh/year^2.

CONCLUSION

A maximization method to allocate recovery energy machines in distribution networks was developed. This methodology was based on Simulated Annealing algorithm. The method enabled to evaluate each transformation in objective function directly. The use of the optimized solution improved the convergence to new optimized solutions for $n+1$ turbines, as well as the search started in the best conditions.

Simulated Annealing algorithm showed good performance when it was compared with the analysis of all possible combinations in the same water system. Small differences only appeared between simulated procedure and all combinations when Ψ_2 was analyzed. The achievement of this small difference justified the use of optimization procedures, withless computational effort.

The combined use of traditional optimization techniques joined to specific software (e.g., Epanet) contributed to develop strategies to improve water management. This improvement will get better efficiency values, causing the upgrade of the sustainability indicators in water systems, contributing to reduce the CO_2 emissions and therefore as well as fighting with the climate change but ensuring quality of the service all the time.

References

Araujo, L. S., Ramos, H., Coelho, S. T., 2006. Pressure Control for Leakage Minimisation in Water Distribution Systems Management. *Water Resour. Manag.* 20, 133–149. doi:10.1007/s11269-006-4635-3.

Castro, A. (2006). *Minicentrales hidroeléctricas*. Instituto Para La Diversificación Y Ahorro de La Energía [*Hydroelectric mini-plants*. Institute for Diversification and Energy Saving]. Madrid, Spain.

Cunha, M., & Ribeiro, L. (2004). Tabu search algorithms for water network optimization. *European Journal of Operational Research*, 157(3), 746-758.

Dougherty, D. E., & Marryott, R. A. (1991). Optimal Groundwater Management: 1. Simulated Annealing. *Water Resources Research*, 27(10), 2493–2508. https://doi.org/10.1029/91WR01468.

Giugni, M., Fontana, N., Ranucci, A., 2014. Optimal Location of PRVs and Turbines in Water Distribution Systems. *J. Water Resour. Plan. Manag.* 140, 06014004. doi:10.1061/(ASCE)WR.1943-5452.0000418.

Kaviani, A., Baghaee, H., Riahy, G., 2009. Optimal Sizing of a stand-alone wind/photovoltaic generation unit using particle swarm optimization. *Simulation* 85, 89–99.

Kirkpatrick S, Gelatt C, Vecchi M. Optimization by simulated annealing. *Science* (80-). 1983;220(4598):671-680.

Metropolis, N., Rosenbluth, A., Rosenbluth, M., Teller, A., & Teller, E. (1953). Simulated annealing. *Journal of Chemical Physics*, 21, 1087–1092.

Pérez-Sánchez, Modesto; Sánchez-Romero, Francisco-Javier; Helena M. Ramos; López Jiménez, Petra Amparo. (2016) Modelling Irrigation Networks for the Quantification of Potential Energy Recovering: *A Case Study. Water*, 6 (8), 1 - 26. 10.3390/w8060234.

Pérez-Sánchez, Modesto; Sánchez-Romero, Francisco-Javier; Ramos, Helena M.; López Jiménez, Petra Amparo. (2017) Optimization Strategy for Improving the Energy Efficiency of Irrigation Systems by Micro Hydropower: *Practical Application. Water,* 1 (9), 10.3390/w9100799.

Pérez-Sánchez, Modesto; Sánchez-Romero, Francisco-Javier; López Jiménez, Petra Amparo; Helena M. Ramos. (2018) PATs selection towards sustainability in irrigation networks: Simulated annealing as a water management tool. *Renewable Energy*, 116 (218), 234-249. 10.1016/j.renene.2017.09.060.

Prats, A. G., Picó, S. G., Alzamora, F. M., Bello, M. Á. J., 2011. *Random Scenarios Generation with Minimum Energy Consumption Model for Sectoring Optimization in Pressurized Irrigation Networks Using a Simulated Annealing Approach.* http://dx.doi.org/10.1061/ (ASCE) IR.1943-4774.0000452.

Ramos, H., Borga, A., 1999. Pumps as turbines: an unconventional solution to energy production. *Urban Water* 1, 261–263. doi:10.1016/S1462-0758(00)00016-9.

Rossman, L. A. (2000). *EPANET* 2: users manual.

Samora, I., Franca, M., Schleiss, A., Ramos, H., 2016a. Simulated Annealing in Optimization of Energy Production in a Water Supply Network. *Water Resour. Manag*. 30, 1533–1547. doi:10.1007/s11269-016-1238-5.

San Bruno, G., Choulot, A., & Denis, V. (2010). Energy recovery in existing infrastructures with small hydropower plants. *Multipurpose Schemes–overview and Examples.* ESHA, SHAPES & Mhylab.

White FM. *Fluid Mechanics*. Sixth edit. McGrau-Hill; 2008.

Zema, D. A., Nicotra, A., Tamburino, V., & Zimbone, S. M. (2016). A simple method to evaluate the technical and economic feasibility of micro hydro power plants in existing irrigation systems. *Renewable Energy*, 85, 498–506.

In: Simulated Annealing
Editors: Alex Scollen et al.
ISBN: 978-1-53613-674-6

Chapter 7

A VARIABLE MARKOV CHAIN LENGTH STRATEGY FOR IMPROVING SIMULATED ANNEALING CONVERGENCE BEHAVIOR: AN EXPERIMENTAL VERIFICATION

F. Javidrad[1,*], PhD, M. Nazari[1] and H. R. Javidrad[2]
[1]Department of Aerospace Engineering,
Aeronautical University of Science and Technology, Tehran, Iran
[2] Department of Mechanical Engineering,
Iran University of Science and Technology, Tehran, Iran

ABSTRACT

The Simulated Annealing (SA) method is a heuristic optimization method inspired from the physical process by which a crystal is formed during slow cooling of melt until the global minimum energy state is reached. The most important feature of SA is perhaps the uphill movement capability using the Metropolis procedure. Although SA is generally known as a reliable method in finding the global optimum in both

* Corresponding Author Email: javidrad.farhad@ssau.ac.ir.

continuous and discrete functions and can produce extremely competitive solution as compared to other heuristic optimization methods, accurate tuning of its parameters to achieve a reasonable performance is crucial. One of these parameters is the Markov chain length (MCL), i.e., the number of iterations at a constant temperature for the system to reach a steady-state condition. In fact, the MCL should be long enough to satisfy a kind of equilibrium condition between the function value and temperature. In the most available SA algorithms, the MCL is determined experimentally and considered constant throughout the simulation. A strategy has previously been proposed that the MCL vary actively according to the optimization function variation. In this regard, temperature decrease is related directly to the function value variation history. One possible relation between these two in a minimization problem is the assumption of a constant temperature in trial solutions with decreasing or constant cost function values and changing temperature in trial solutions with increasing cost function. Because the temperature in this algorithm is changing variably with the objective function, an accelerated approach to the global optimum is expected due to the faster temperature decrease at high temperatures.

In this chapter, performance of this variable MCL method in unconstrained optimization problems has been studied experimentally. For this purpose, eighteen benchmark optimization functions with different dimensions were employed to compare performance and convergence behavior of this algorithm to the standard SA algorithm with different Markov chain lengths. Three different criteria were used for this comparison. The results show that the variable MCL algorithm can give rise to an improvement in convergence speed.

Keywords: simulated annealing, unconstrained optimization, Markov chain length, cooling process

INTRODUCTION

New design techniques based on stochastic search optimization methods have emerged in various engineering disciplines such as aeronautical, structural and mechanical engineering. Commonly, stochastic search techniques make use of the ideas inspired from nature with some heuristic manners to find the global optimum of a problem and do not suffer the discrepancies encountered in most of the deterministic optimization methods. Deterministic optimization methods usually require preconditions such as continuity and differentiability of the objective function which are not met in many practical problems [1]. On the other hand, stochastic search optimization methods can successfully be applied to problems including non-smooth or discontinuous functions of continuous, discrete and integer variables. These methods are easy to implement because they require only function values to find an optimum. Besides, stochastic methods are usually capable of jumping out of the local minima to approach the global or near global optimum solution [2]. However, heuristic methods have a number of drawbacks. These methods may be inefficient in terms of computational effort because they do need thorough exploration across the feasible area. In addition, there are some factors that must be set experimentally to assure that the stochastic method approaches to the global optimum point.

The stochastic search methods are usually classified in three groups: Monte Carlo based probabilistic search, evolutionary algorithms and swarm intelligence. To date, a number of stochastic heuristic or meta-heuristic search methods from all three so-called groups have been developed. Simulated annealing (SA) and tabu search (TS) from the probabilistic search category, genetic algorithms (GA) from the category of evolutionary algorithms and particle swarm optimization (PSO) and ant colony optimization (ACO) from the swarm intelligence category are the methods that have been widely used to address optimization issues [3]. Amongst all these methods, SA has accelerated gaining attention by scholars and researchers because of its simple concepts, good speed and ease of implementation.

Simulated annealing was developed in 1983 by Kirkpatrick [4] to deal with highly nonlinear problems. SA approaches the global optimization problem similarly to using a bouncing ball that can bounce over peaks from valley to valley. The method usually begins with high energy which enables very high bounces, i.e., it bounce over any peaks in search domain. Bouncing energy is illustrated by a parameter called temperature, analogous to the oscillation energy of molecules of a heated metal. As the temperature drops the bouncing is limited, so that it can settle to become trapped in relatively small ranges of valleys. Figure 1 shows a typical variation of an objective function value in a minimization problem by SA procedure [5].

In SA method an acceptance distribution is defined to accept or reject a trial solution. This acceptance distribution is a function of the difference between the trial position function value and the last saved position and the temperature. This acceptance distribution actually decides probabilistically whether to stay in a new lower energy state or to bounce out of it. It has been proved that by using extremely slow rate of cooling of the temperature, SA is able to find the global optimum point [6]. However, this requires infinite time. Fast annealing and very fast simulated re-annealing (VFSR) or variable simulated annealing (ASA) are each in turn exponentially faster and can overcome this problem [7].

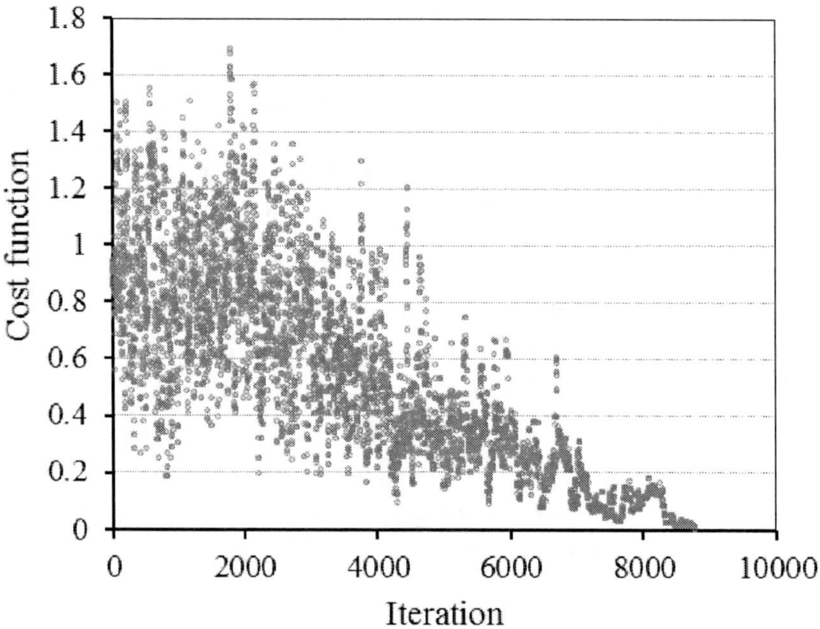

Figure 1. A typical variation of an objective function value in a minimization problem using SA method [5].

The remainder of this chapter is structured as follows: Section 2 gives a brief description of the SA method. In Section 3 the variable MCL algorithm is described. Experimental work and the evaluation of the variable MCL algorithm through benchmark function are given in Section 4 and Section 5, respectively. Finally, Section 6 and Section 7 respectively presents discussions and concluding remarks.

BASICS OF THE SA METHOD

The SA method simulates the physical process by which a crystal is formed during slow cooling of melt until the global minimum energy state is reached. SA explores entire domain of the objective function and tries to optimize it while moving both uphill and downhill. This is definitely the major advantage of SA method which cause to avoid becoming trapped in local minima. At the heart of the SA algorithm is the Metropolis Monte Carlo procedure, which was introduced to provide an efficient simulation of a collection of atoms in equilibrium at a given temperature. The Metropolis procedure allows moves to states with a higher objective function value, while the standard Monte Carlo algorithm prevents changes of state that increase the objective function. In this regard, SA can be considered as a variant of Monte Carlo search technique.

In SA algorithm, a candidate move from the current solution, X_i, to a next trial state, X_{i+1}, in the neighborhood of that point is accepted with probability

$$P = \begin{cases} 1 & \textit{if} \quad f(X_{i+1}) \le f(X_i) \\ e^{-(f(X_{i+1})-f(X_i))/T} & \textit{Otherwise} \end{cases} \tag{1}$$

where T is a control parameter, called temperature. The temperature in SA algorithm is sequentially lowered until the system freezes, i.e., no further changes occur. At each temperature, the simulation should proceed long enough for the system to reach a steady-state condition. The sequence of temperatures and the method to reach a quasi-equilibrium condition at each temperature is known as a cooling process, which is comparable to an annealing schedule normally applied in annealing of metals. The cooling process is actually an attempt to approximate the theoretical algorithm in light of a finite amount of computation time. Normally, there are four parameters to be specified for any cooling process implementation:

a) a generator of random changes in solutions,
b) a means of evaluating the problem function values,
c) an annealing schedule - an initial temperature and rules for lowering it as the search progresses and
d) a stop criterion for terminating problem execution.

Steps for a SA procedure can be describes in a Pseudo-code as below:

Procedure SA

S ← GenerateInitialState
// {define an initial vector of design variables}
T ← SetInitialTemperature, T_0
n_{max} ← MaxAllowableIter
MCL ← MarkovChainLength
R ← TemperatureDecrementRatio
n ← NumberVariables
Stopflag=.false.
for *i* from 1 to $n_{\max}$
for *j* from 1 to *MCL*
for *k* from 1 to *n*
S' ← PickRandomState
// {select a vector of design variables}
If f(S')=< f(S)
// {compare the cost function value}
S ← S' {accepting a move}
else
r ← Random(0,1)
if r < exp((f(S)-f(S'))/T) then S ← S'
// {accepting a move}

endif
If Stopflag=.true. then Return {check for the stop criteria}
endfor
endfor
T ← T*R {modify temperature}
endfor
end

In the above procedure, there are some parameters that must be defined before the SA procedure running. These parameters are explained below.

1) The initial temperature value

The selection of the initial temperature can affect the quality and speed of solution. If the initial temperature is too low, the step length may be too small to catch the optimum point. If too high, then the step length could be too large and an excessively large area is searched. A proper value of T_0 will clearly depend on the objective function behavior and, hence, be problem-specific. An appropriate initial temperature, T_0, is one that results in a high acceptance probability of change, which increases the objective function.

2) The decrement rule for updating the temperature value

The choices made for each parameter of the temperature updating rule differentiate one implementation of the SA algorithm from another. There are essentially two types of decrement rules defining two categories under which most SA cooling process fall: constant and variable. In constant decrement rules, temperature change occurs after a fixed number of iterations and are generally independent of the objective function change, usually of the form

$$T_i = R^i T_0 \tag{2}$$

where R is a positive constant less than one. This constant is usually taken in the range of 0.8 – 0.999. It is noted that choosing relatively small values for *R* makes a rapid temperature decrement and therefore can cause the algorithm get stuck into a local minimum point. Variable decrement rules, on the other hand, vary dynamically the proportional size of the temperature decrements over the course of the algorithm [8].

3) The length of trial solutions at each temperature, i.e., the length of Markov chain

In SA, we start by constructing a sequence of temperatures T_1, T_2 and so on. At each step of this sequence, moving to neighboring positions are taking place for MCL-times (see the SA pseudo-code). Such a stochastic construction of a sequence is called a Markov chain. It can be shown for very long chain lengths the candidate solution will visit every possible positions. Such Markov chains are closely inspired by physical models where equilibrium of molecules movement in a specified temperature has been taken place in annealing process. There are theoretical results on how slowly the annealing has to be done to be sure that one reaches the globally optimum solution, but in practice the running times are enormous. There are a few researches in literature concerning the effect of the MCL on the solution quality and speed, e.g., [9].

4) Stop criterion

Stop criterion is actually the conditions under which execution terminates.

The Variable MCL SA Method

As mentioned above, at each temperature, the search process should proceed long enough for the system to reach a steady-state condition. It means that the MCL should satisfy a kind of equilibrium condition between the cost function value and the temperature. In the most available SA algorithms, the MCL is determined experimentally and considered constant throughout the simulation. In the variable MCL algorithm, which is introduced in [5], the MCL is changing actively according to the cost function change. In this regard, temperature decrease performed according to Eq. 2 whenever the cost function increases. Therefore, in trial solutions with decreasing or constant cost function value, temperature remains unchanged. Because the temperature in this algorithm is changing adaptively with the cost function, an accelerated approach to the global optimum is expected due to the faster temperature decrease at high temperatures. The pseudo-code of this SA algorithm is given below.

```
Procedure VariableMCLSA

S ← GenerateInitialState
// {define an initial vector of design variables}
T ← SetInitialTemperature, T0
R ← TemperatureDecrementRatio
n ← NumberVariables
Stopflag=.false.
for i from 1 to nmax
for k from 1 to n
S' ← PickRandomState
// {select a vector of design variables}
If f(S')=< f(S)
// {compare the cost function value}
S ← S' {accepting a move}
else
r ← Random(0,1)
if r < exp((f(S)-f(S'))/T) then S ← S'
// {accepting a move}
endif
endfor
If Stopflag = .true. then Return {check for the stop criteria}
```

```
If f(S') > f(S)
T ← T*R {modify temperature}
endif
endfor
end
```

EXPERIMENTAL WORK

In this book chapter, we aimed at finding that how this variable temperature change affects the global solution convergence properties in unconstrained optimization problems. To find the answer of this question, we performed extensive numerical experiments. A number of unconstrained optimization functions with different dimensions have been employed as benchmarks for global minimization and have been solved by the standard SA and the modified one. In the standard SA procedure, different MCLs have been applied. Because of the stochastic nature of the SA optimization method, the solution for each function were independently repeated 100 times. In all runs, the initial temperature T_0 and R were set to 1 and 0.95, respectively.

The initial candidate solution is selected from 50 random positions. In this regard, the best position of 50 random positions are defined as the start point.

The successful runs are defined as the computer runs where the cost function getting values less than a tolerance value, 1×10^{-3}. The failed runs, however, are those where at least one of the following criteria is satisfied:

- number of cost function evaluation exceeding 2×10^4,
- temperature reaching to 1×10^{-4}, and
- cost function values being unchanged or getting worse for 100 successive iterations.

The effectiveness and efficiency of the solution in stochastic procedures can generally be defined in terms of the percentage of successful runs, convergence speed and the maximum effective trial function evaluations. The latter is, in fact, a true measure of efficiency considering both successful and failed runs in a prescribed number of runs. In this regard, three parameters have been used in this study to compare the performance of the variable MCL SA method. These parameters are:

- Stability of solution: the percentage of successful solutions.
- Average trial: the average number of function evaluations in successful runs. This term is calculated by averaging number of function evaluations in successful solutions achieved in a limited number of independent runs, say 100.
- Total average trial: the average number of function evaluations considering all successful and failed runs, as determined according to Eq. 10. In fact, the combined effect of function evaluations for successful runs and stability is included in the total average function evaluations. This equation has been suggested in [5]:

$$(Ave)_T = (Ave)_s + \frac{Fail}{Success}(Ave)_f \tag{3}$$

where $(\text{Ave})_T$, $(\text{Ave})_s$, $(\text{Ave})_f$, *Fail* and *Success* denote the total average trial function evaluations, the average of trial function evaluations for successful runs, the average number of function evaluations used up for failed runs, the numbers of failed and successful runs in total *n* runs, respectively. The computational cost actually consists of function evaluations used up for both successful and failed runs. Therefore, the fraction of function evaluations of failed runs attributed to every successful run is added to the average function evaluations for successful runs to give the actual computational effort for a limited number of runs. The term $Fail.(Ave)_f/Success$ is the fraction of total failed function evaluations allocated to every successful run. It is obvious that when stability percentage is 100, the term $Fail.(Ave)_f/Success$ is zero and therefore, the average trial and the total average trial are equal.

Experimental Results

Characteristics of Test Functions

In general, the goal of any global optimization problem is to find the best possible solutions x^* from a set X according to a set of criteria $F = \{f_1, f_2, \cdots f_n\}$. These criteria which are called objective functions are frequently expressed in the form of a mathematical functions or tabulated values. An objective function is usually defined in a set D ($f : D \subset \Re^n \to \Re$) which is referred to as the set of feasible points in a search space. In the case of optimizing a single objective function, an optimum is either its maximum or minimum. The global optimization problems are often defined as minimization problems, however, these problems can be easily converted to maximization problems by negating *f*.

The true optimal solution of an optimization problem may be a set of $x^* \in D$ of all optimal points in *D*, rather than a single minimum or maximum value in some cases. There could be multiple, even an infinite number of optimal solutions, depending on the domain of the search space. Finding globally optimal solutions or at least sub-optimal solutions are the output of the global optimization algorithm. The objective functions are frequently classified as their properties such as dimensionality, continuity, convexity, modality and separability. Basic definitions of these properties are given.

1) Dimensionality: The difficulty of a problem generally increases with its dimensionality. As the number of parameters or dimension increases, the search space also increases exponentially. For highly nonlinear problems, this dimensionality may be a significant barrier for almost all optimization algorithms.
2) Continuity: An *n* variable real-valued function $f(x_1, x_2, \cdots, x_n)$ is said to be continuous at the point $(x_1, x_2, \cdots, x_n) \in D$ if $\lim_{(x_1,\, x_2, \cdots, x_n) \to (a_1,\, a_2, \cdots, a_n)} f(x_1, x_2, \cdots, x_n) = f(a_1, a_2, \cdots, a_n)$. A function (as a

whole) is said to be continuous, if it is continuous at every point within its domain of definition.

3) Convexity: A real-valued function defined on an n-dimensional interval is called convex (or convex downward or concave upward) if the line segment between any two points on the graph of the function lies above or on the graph, in a generalized Euclidean space (or more generally a vector space) of at least two dimensions. Equivalently, a function is convex if its domain is a convex set in $\Re^n$ and for all x_1 and x_2 in domain D and $\alpha \in [0,1]$

$$f(\alpha x_1 + (1-\alpha)x_2) \leq \alpha f(x_1) + (1-\alpha) f(x_2) \tag{4}$$

4) Modality: The number of optimum (minima or maxima) present in the search space defines the modality of a function. A function is then multimodal if it has two or more local optima. Modality is important for optimization of functions because for multimodal functions, the optimization algorithm may converge to a local optima during the search process, and thereby fail to find the global optima. Thus, the search process may slow down and it is difficult to find true optimal solution for multimodal functions.
5) Separability: The separability of a function is a measure of the difficulty in solving the optimization problem using evolutionary algorithms. If the objective function variables are independent of each other, then the objective functions can be decomposed into sub-objective functions. Then, each of these sub-objectives involves only one variable, while treating all the others as constant. Thus, a function of n variables $f(x_1, x_2, \cdots, x_n)$ is said to be separable if it can be expressed in terms of n functions $f_1, f_2, \cdots, f_n$, such that

$$f(x_1, x_2, \cdots, x_n) = \sum_{i=1}^{n} f_i(x_i) \tag{5}$$

In general, separable functions are relatively easy to solve, when compared with their inseparable counterpart. If all the parameters or variables are independent, then a sequence of n-independent optimization processes can be performed. As a result, each variable or parameter can be optimized independently.

Optimization of Benchmark Functions

1. Two-Dimensional Shubert Function

$$f(x_1,x_2)=[\sum_{j=1}^{5} j\cos((j+1)x_1+j)]\times[\sum_{j=1}^{5} j\cos((j+1)x_2+j)], \quad x_1,x_2\in[-10,10] \tag{6}$$

The Shubert function is a multi-modal function arises in signal processing applications. The 2-dimensional Shubert function is continuous, differentiable and separable and has several local minima and 18 global minima [10]. The global optimum solution is -186.730. This function exhibits a complex landscape as shown in Figure 2.

The problem has been solved by the fixed MCL SA method considering three different MCLs and the variable MCL SA method. As the range of MCL values are usually taken between 20 and 80, we selected three values for MCL in this range. Table 1 gives the obtained results. It is seen that although the variable MCL method has equal success rate with MCL = 80, it presents much better convergence rate. The total average of function evaluations, $(Ave)_T$, for 100 independent runs obtained according to Eq. 3 for variable MCL SA strategy is only about 40% of the fixed MCL = 80. Figure 3 shows typical convergence curves for various MCLs and that of the variable MCL SA method.

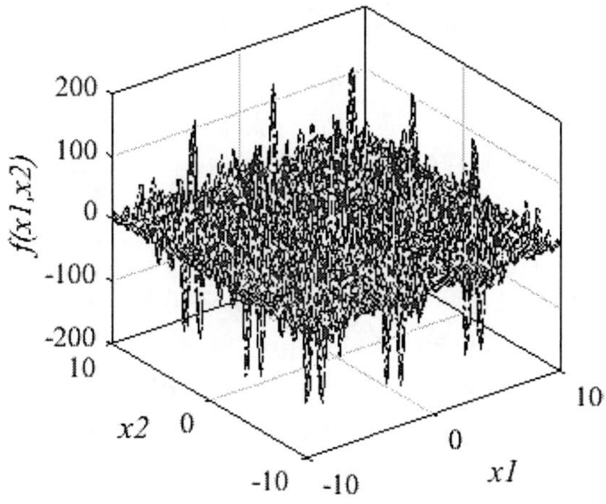

Figure 2. The Shubert function.

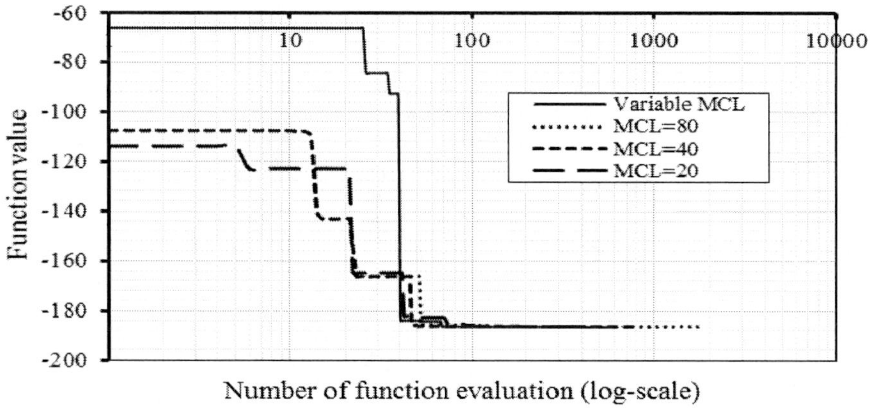

Figure 3. Typical convergence curves for optimization of the Shubert function.

Table 1. Shubert function optimization results and statistics

MCL	20	40	80	Variable
Success rate (%)	79	85	86	86
Average function evaluations	501.785	842.424	1606.31	652.453
Total average	1913.05	2731.72	4834.59	1831.76
Best values	-186.7309	-186.7309	-186.7309	-186.7308
Worst values	-38.29452	-49.51861	-49.51862	-49.51862
Mean	-169.9890	-174.4411	-176.4561	-176.4563
Variance	36.43042	31.64279	27.13698	27.13701

2. Two-dimensional Holder's Table Function

$$f(x_1,x_2) = -\left|\sin x_1 \cos x_2 \exp\left(\left|1-\frac{\sqrt{x_1^2+x_2^2}}{\pi}\right|\right)\right|, \qquad x_1,x_2 \in [-10,10] \qquad (7)$$

The 2D Holder's Table function is continuous, not convex, non-differentiable, non-separable and multimodal [10]. It has many local minima, with four global minima. The global optimum solution is −19.20850 at (8.05502, 9.66459), (8.05502, -9.66459), (-8.05502, 9.66459), (-8.05502, -9.66459). The 2D Holder's Table function is depicted in Figure 4.

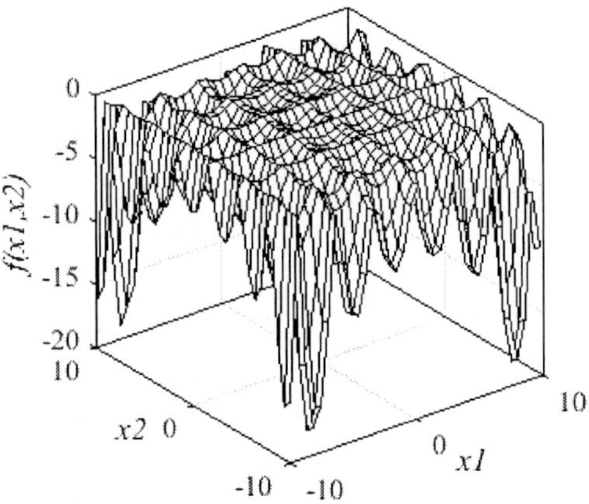

Figure 4. The 2D Holder's Table function.

Table 2. Holder's Table function optimization results and statistics

MCL	20	40	80	Variable
Success rate (%)	95	97	100	98
Average function evaluations	861.968	1196.61	1860.14	592.5
Total average	580.179	861.041	1860.14	559.643
Best values	-19.2085	-19.2085	-19.2085	-19.20847
Worst values	-8.09512	-8.09512	-19.207	-8.095123
Mean	-18.7116	-18.9041	-19.2078	-18.98546
Variance	2.187042	1.743538	0.000434	1.555762

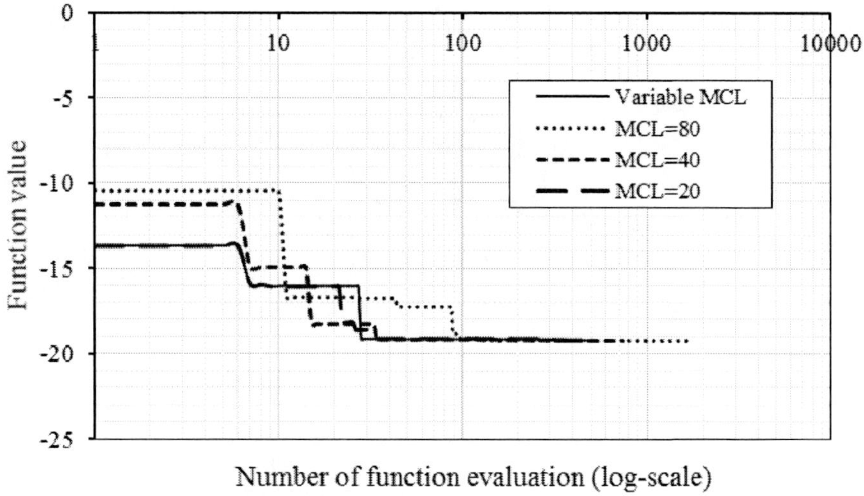

Figure 5. Typical optimization convergence curves of the Holder's Table function.

Table 2 presents the data obtained for three different MCL values and the proposed variable MCL SA method. The obtained results show that the variable MCL SA has 2% less success rate as compared to fixed MCL = 80 but it needs only one-third function evaluations. On the other hand, comparison between the variable MCL SA method with the fixed MCL = 20 shows that the total averages are close to each other but the variable MCL SA has 3% more success rate. Typical convergence curves for this function are exhibited in Figure 5.

3. Two-Dimensional Bohachevsky Function

$$f(x_1, x_2) = x_1^2 + 2x_2^2 - 0.3\cos(3\pi x_1 + 4\pi x_2) + 0.3, \qquad x_1, x_2 \in [-100,100] \tag{8}$$

The 2D Bohachevsky function is a bowl-shaped function as shown in Figure 6. The function is continuous, differentiable, non-separable and has one global minimum at (0,0) with the global minimum value of 0. [10].

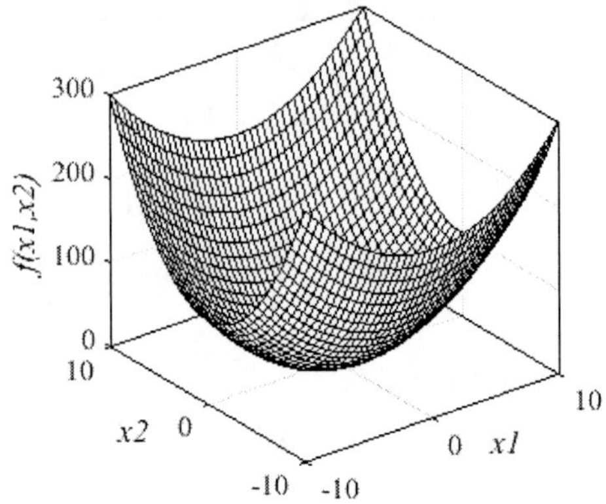

Figure 6. The Bohachevsky's function.

Table 3. Bohachevsky function optimization results and statistics

MCL	20	40	80	Variable
Success rate (%)	51	92	100	97
Average function evaluations	853.843	1619.23	3360.11	774.113
Total average	6090.12	2562.71	3360.11	824.320
Best values	8.85E-05	1.73E-05	2.98E-07	1.41561E-05
Worst values	0.228171	0.006538	0.000997	0.072454482
Mean	0.012034	0.000766	0.000494	0.001754042
Variance	0.031944	0.000983	0.000296	0.007992668

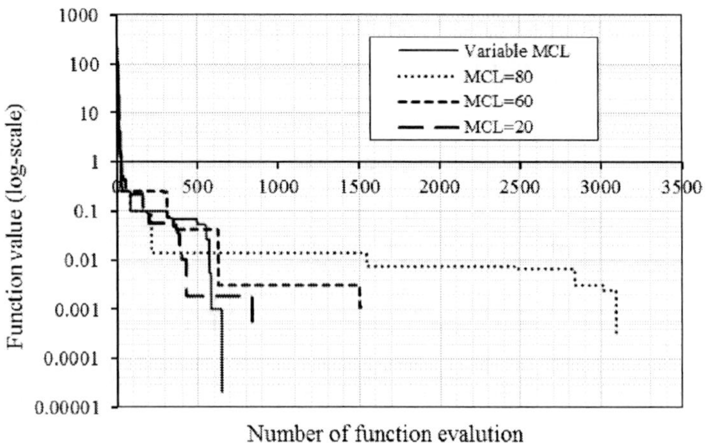

Figure 7. Typical optimization convergence curves of the Bohachevsky function.

Table 3 shows the results obtained for three different values of MCL together with those of the variable MCL SA method. The results show that although the stability of the proposed variable MCL method is slightly lower than the fixed MCL = 80, it is more than four times faster. Typical convergence curves for this function are exhibited in Figure 7.

4. Two-Dimensional Beale Function

$$f(x_1,x_2) = (1.5 - x_1 + x_1x_2)^2 + (2.25 - x_1 + x_1x_2^2)^2 + (2.625 - x_1 + x_1x_2^3)^2 \quad (9)$$

The Beale function is continuous, differentiable, non-separable and unimodal function with sharp peaks at the corners of the input domain as shown in Figure 8. The function is usually evaluated in the domain of $x_1, x_2 \in [-4.5, 4.5]$. It has a global minimum value of 0. located at (3,0.5) [10].

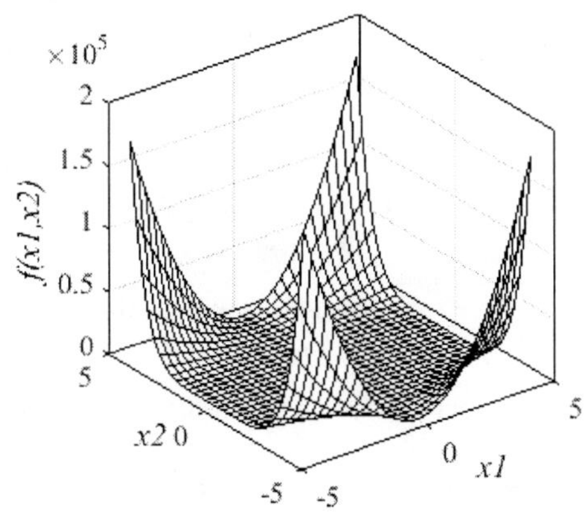

Figure 8. The Beale function.

Table 4 shows the results obtained for this function. Again, three different MCLs have been used to evaluate the performance of the variable MCL SA method. The results show that the variable MCL SA method has 78% success rate which is the same as MCL = 40 and 80. However, the total average of the variable MCL SA is at least 30% of the others. Typical convergence curves for this function are presented in Figure 9.

Table 4. Beale function optimization results and statistics

MCL	20	40	80	Variable
Success rate (%)	61	78	78	78
Average function evaluations	629.672	862.769	1779.32	610.82
Total average	4114.1	3923.03	7420.63	1204.41
Best values	1.83E-05	6.04E-06	1.18E-05	2.42E-05
Worst values	2.051104	0.762069	0.76207	0.762072
Mean	0.217247	0.168054	0.16805	0.168092
Variance	0.43444	0.315473	0.315475	0.315453

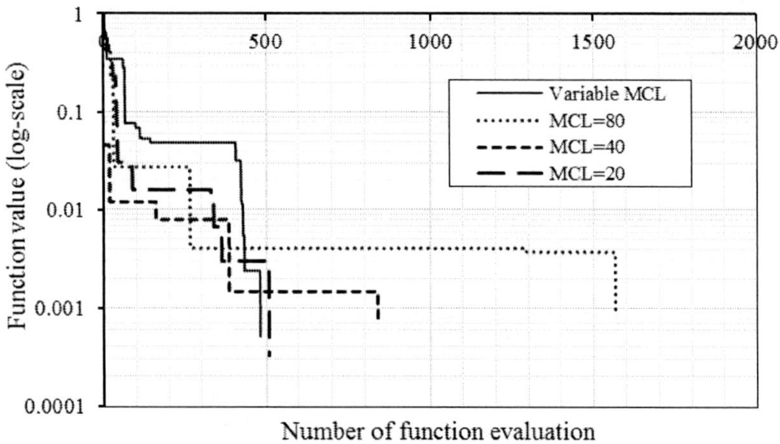

Figure 9. Typical optimization convergence curves of the Beale function.

5. Two-Dimensional Styblinski-Tang Function

$$f(x_1,x_2) = 0.5\sum_{j=1}^{2}(x_j^4 - 16x_j^2 + 5x_j), \qquad x_1, x_2 \in [-5, 5] \tag{10}$$

The non-convex Styblinski-Tang function has four local minima, with the global minimum located at $x_1 = x_2 = -2.903534$, and one maximum located close to the origin [11]. The minimum value of the function is -78.332. The function is shown in Figure 10.

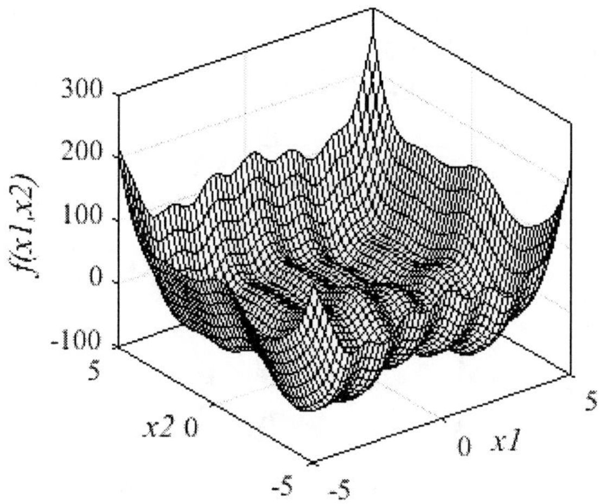

Figure 10. The Styblinski-Tang function.

Table 5 gives the results obtained by the fixed and variable MCL SA methods. The results show that the variable MCL strategy has more than three times convergence speed as compared to that of the fixed MCL for the same success rate. The best and worst calculated global optimum values which are related to the quality of solution remained unchanged in all four solutions. Typical convergence curves for this function are presented in Figure 11.

Table 5. Styblinski-Tang function optimization results and statistics

MCL	20	40	80	Variable
Success rate (%)	89	93	96	96
Average function evaluations	619.573	846.387	1735.76	543.365
Total average	1275.64	1632.09	2569.14	838.781
Best values	-78.3323	-78.3323	-78.3323	-78.3323
Worst values	-64.1956	-64.1956	-64.1956	-64.1956
Mean	-76.9181	-77.3422	-77.7662	-77.7663
Variance	4.240819	3.606778	2.770088	2.770095

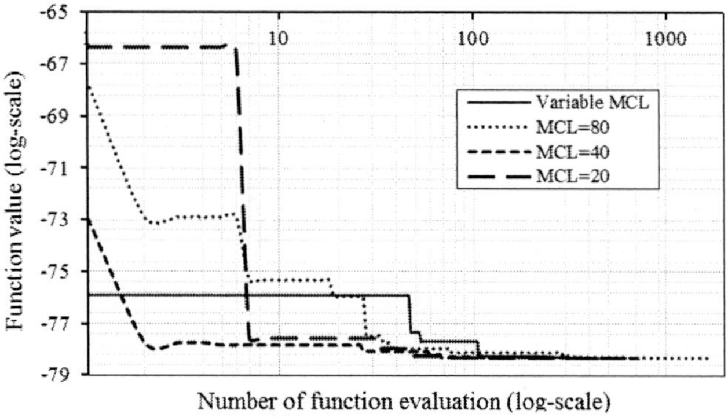

Figure 11. Typical optimization convergence curves of the Styblinski-Tang function.

Table 6. Styblinski-Tang function with noisy local minima optimization results and statistics

MCL	20	40	80	Variable
Success rate (%)	76	86	90	90
Average function evaluations	661.921	997.407	2213.62	573.956
Total average	2357.18	2714.38	4408.94	1054.18
Best values	-83.6862	-83.6862	-83.6862	-83.6862
Worst values	-63.3769	-69.8851	-69.8851	-69.8851
Mean	-80.7695	-81.9775	-82.44	-82.44
Variance	5.421238	4.294434	3.773049	3.773042

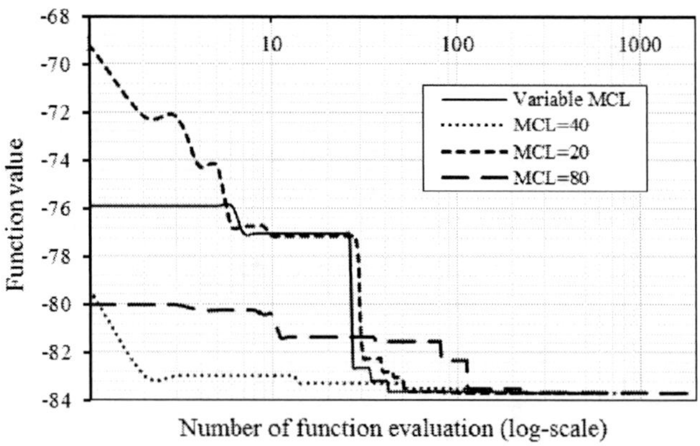

Figure 12. Typical optimization convergence curves of the Styblinski-Tang function with noisy local minima.

As an extension of this case study, a mathematical term has been added to the Styblinski-Tang function to include more local minima (Eq. 11). This function, which is introduced in [11], has been used to test the convergence behavior of the optimization methods in the presence of a large number of noisy local minima. The global minimum of this function is -83.6862. The results, as given in Table 6, still showed superior performance of the variable MCL SA method over the fixed MCL SA method. The typical convergence curves are presented in Figure 12.

$$f(x_1,x_2)=0.5\sum_{j=1}^{2}(x_j^4-16x_j^2+5x_j)-10\cos(4(x_1-2.093534))\cos(2(x_2-2.093534))$$
$$x_1,x_2\in[-5,5] \qquad (11)$$

To further investigate the effectiveness of the variable MCL SA method in finding the global minimum of this function, the 5-dimensional Styblinski-Tang function, as suggested in [11], (Eq. 12) has also been treated. The global minimum of this function is -78.332. The optimization results, as presented in Table 7 and Figure 13, again confirmed the good performance of the variable MCL SA method.

$$f(x_1,\ldots,x_5)=0.5\sum_{j=1}^{2}(x_j^4-16x_j^2+5x_j)+\sum_{j=3}^{5}(x_j-1)^2,\quad x_j\in[-5,5]\quad j=1,\ldots,5 \qquad (12)$$

Table 7. 5D Styblinski-Tang function optimization results and statistics

MCL	50	75	100	Variable
Success rate (%)	65	77	79	79
Average function evaluations	6148.95	7025.47	9341.14	1578.14
Total average	13445.1	12999.8	14657.9	3454.72
Best values	-78.3321	-78.332	-78.3322	-78.3321
Worst values	-64.1	-64.1956	-64.1956	-64.1956
Mean	-74.4941	-75.0801	-75.3628	-75.3628
Variance	6.268768	5.948781	5.757617	5.757599

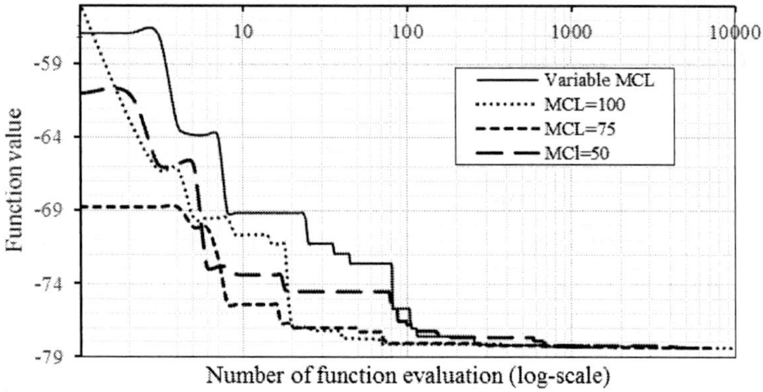

Figure 13. Typical optimization convergence curves of the 5D Styblinski-Tang function.

6. Two-Dimensional Table 3/Carrom Table Function

$$f(x_1,x_2) = -\frac{(\cos(x_1)\cos(x_2)\exp\left|1-\frac{(x_1^2+x_2^2)^{0.5}}{\pi}\right|)^2}{30}, \quad x_1,x_2 \in [-10,10] \tag{13}$$

The Table 3/Carrom Table function is continuous, differentiable and separable function. The function is multi-modal with four global minimum points at $x_i \approx \pm 9.646157$. The global minimum function is –24.1568155. The function has very complex landscape as shown in Figure 14.

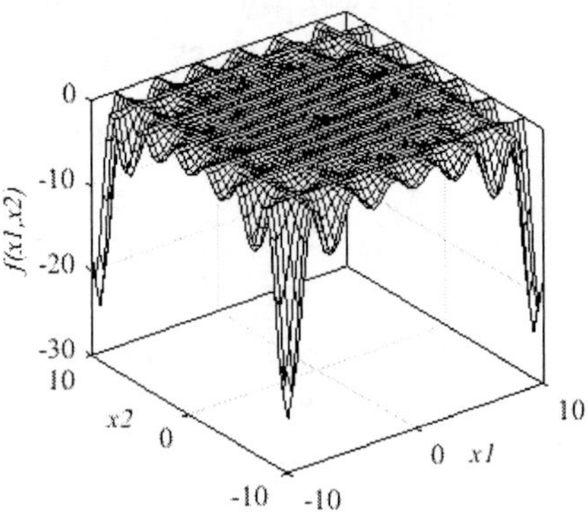

Figure 14. The Table 3/Carrom Table function.

Table 8. The Table 3/Carrom Table function optimization results and statistics

MCL	20	40	80	Variable
Success rate (%)	91	100	100	100
Average function evaluations	492.011	757.47	1686.89	557.23
Total average	1024.43	757.47	1686.89	557.23
Best values	-24.1568	-24.1568	-24.1568	-24.1568
Worst values	-2.63045	-24.155	-24.155	-24.155
Mean	-23.0707	-24.1558	-24.1559	-24.1559
Variance	4.312243	0.000551	0.000543	0.000485

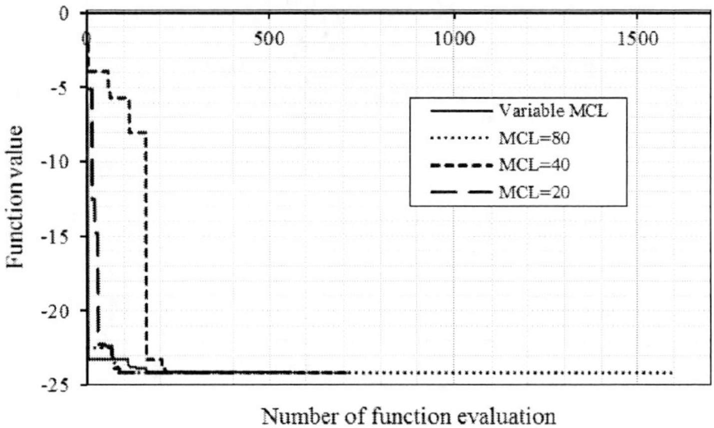

Figure 15. Typical optimization convergence curves of the Table 3/Carrom Table function.

Results of the analysis with fixed MCLs and the variable MCL SA methods are presented in Table 8. It can be seen that the variable MCL SA method needs only 70% function evaluations of the fixed MCL SA for the same success rate and the same solution quality. Typical convergence curves for this function are exhibited in Figure 15.

7. Two-Dimensional Schaffer F6 Function

$$f(x_1,x_2)=0.5+(\sin^2(x_1^2-x_2^2)-0.5)/(1+0.001(x_1^2+x_2^2))^2,\quad x_1,x_2\in[-100,100] \tag{14}$$

The Schaffer F6 function is a continuous, differentiable, non-separable and multimodal function [10]. The difficulty in this function is that the size of the potential maxima that need to be overcome to get to a minimum increases the closer one gets to the global minimum. The global minimum of the function is $x_1 = x_2 = 0$. with the minimum function value of 0. The Schaffer F6 function is depicted in Figure 16.

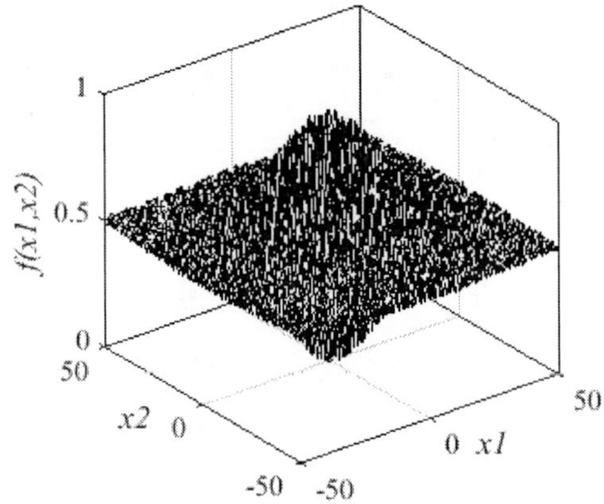

Figure 16. The Schaffer F6 function.

Table 9. The Schaffer F6 function optimization results and statistics

Markov chain length	20	40	80	Variable
Success rate (%)	4	21	47	32
Average function evaluations	238.750	1828.76	4325.02	519.562
Total average	131039	42645.4	26879.3	4460.81
Best values	9.3E-06	3.13E-06	5.36E-06	5.36E-06
Worst values	0.324945	0.101236	0.024212	0.039843
Mean	0.05986	0.013371	0.003952	0.005468
Variance	0.064058	0.015403	0.00507	0.006596

Table 9 presents the results obtained for three different values of MCL together with those of the proposed method. The results show that although the success rate of the proposed variable MCL method is lower than the fixed MCL = 80, it is more than eight times faster. Typical convergence curves for this function are exhibited in Figure 17.

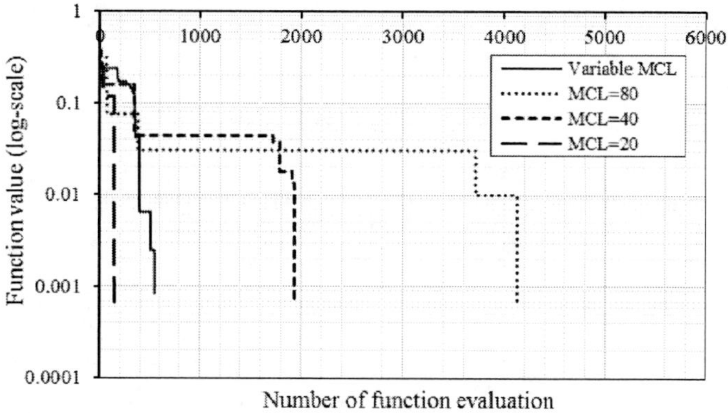

Figure 17. Typical optimization convergence curves of the Schaffer F6 function.

8. Two-Dimensional W/Wavy Function

$$f(x_1, x_2) = 1 - 0.5\sum_{j=1}^{2}\cos(10x_j)\exp(-x_j^2/2), \quad x_1, x_2 \in [-\pi, \pi] \tag{15}$$

The W/Wavy function is a continuous, differentiable, non-separable and multimodal function [10]. The global minimum is located at $x_1 = x_2 = 0$. This function has 121 local minima with the global minimum function value of 0. The function is shown in Figure 18.

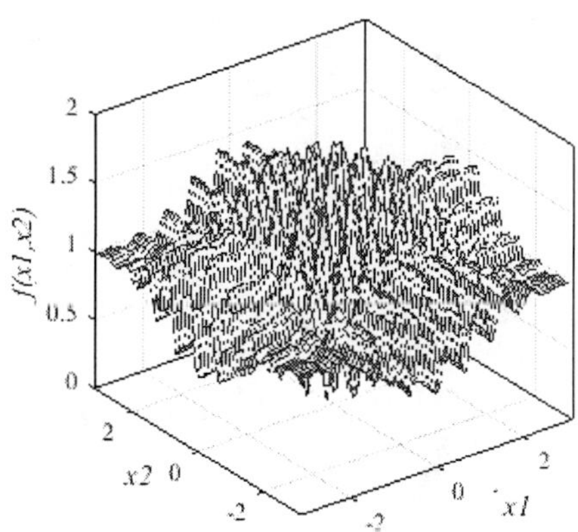

Figure 18. The W/Wavy function.

Table 10. The W/Wavy function optimization results and statistics

MCL	20	40	80	Variable
Success rate (%)	24	25	63	77
Average function evaluations	789.958	1931.44	4113.43	572.584
Total average	18048.3	34481.4	15860	1049.86
Best values	9.3E-06	1.26E-05	2.62E-06	2.36E-05
Worst values	0.324945	0.108306	0.049415	0.088763
Mean	0.05986	0.020064	0.004273	0.009073
Variance	0.064058	0.024883	0.008692	0.021806

The results of three different values of MCL and the proposed variable MCL SA have been given in Table 10. The results show that the variable MCL SA has higher success rate compared to the constant MCLs together with a much lower computational cost. Typical convergence curves for this function are shown in Figure 19.

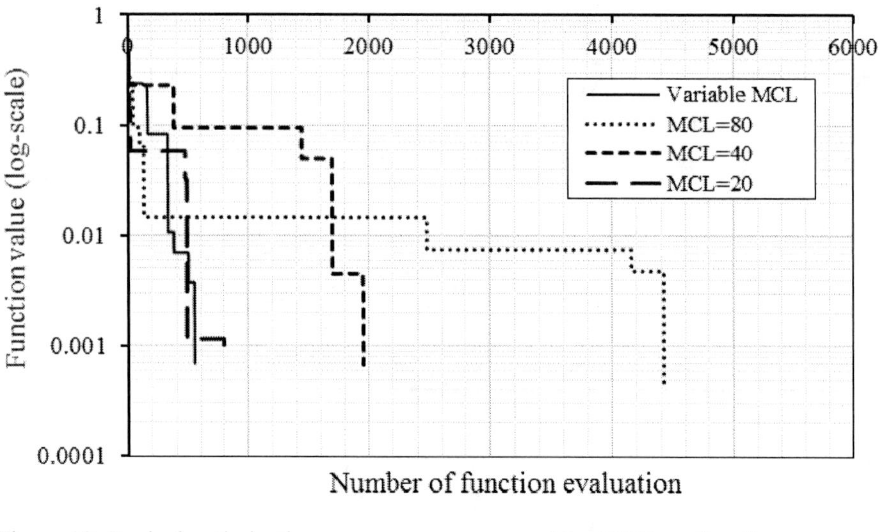

Figure 19. Typical optimization convergence curves of the W/Wavy function.

9. Two-Dimensional Branin RCOS Function

$$f(x_1, x_2) = (x_2 - (5/4\pi^2)x_1^2 + (5/\pi)x_1 - 6)^2 + 10(1 - (1/8\pi))\cos x_1 + 10,$$
$$x_1 \in [-5, 10], \quad x_2 \in [0, 15] \tag{16}$$

The Branin RCOS function is a continuous, differentiable, non-separable and multimodal function [10]. The global minimum is located at $x_1 = x_2 = 0$. The function has three global minima at (-π, 12.275), (π, 2.275), (3π, 2.425) with the global function value of 0.3978873. The Branin RCOS function, as shown in Figure 20, has not local minima.

Results of the analysis with three different fixed MCLs and the variable MCL strategy are given in Table 11. The obtained results show that the variable MCL SA method needs less than one-third function evaluations of the fixed MCL = 40 for the same 100% success rate. The variable MCL SA method also presented better solution quality compared to the fixed MCL SA solution procedure. Typical convergence curves for this function are presented in Figure 21.

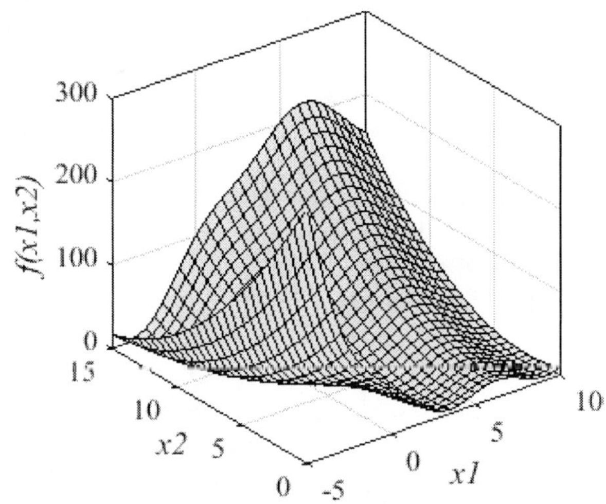

Figure 20. The Branin RCOS function.

Table 11. The Branin RCOS function optimization results and statistics

MCL	20	40	80	Variable
Success rate (%)	97	100	100	100
Average function evaluations	1378.43	2263.23	4894.55	751.83
Total average	1546.99	2263.23	4894.55	751.83
Best values	0.397891	0.397892	0.397888	0.397888
Worst values	0.475901	0.397999	0.39799	0.39799
Mean	0.39878	0.397945	0.397944	0.397944
Variance	0.007768	3.09E-05	3.31E-05	3.31E-05

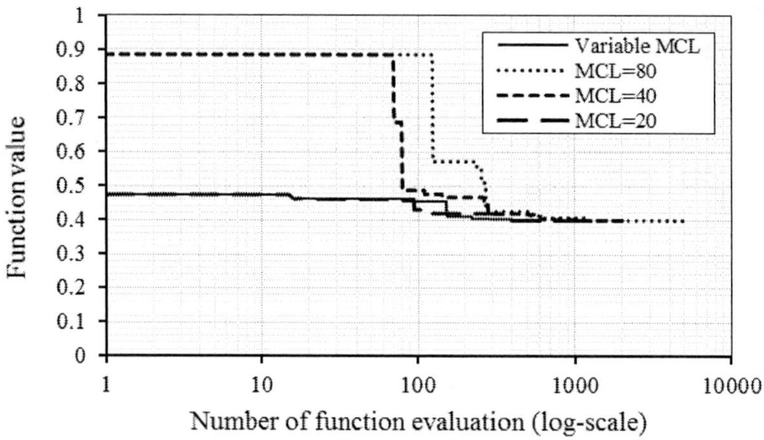

Figure 21. Typical optimization convergence curves of the Branin RCOS function.

10. Four-Dimensional Rastrigin Function

$$f(x_1,\ldots,x_4)=\sum_{j=1}^{4}(x_j^2-10\cos(2\pi x_j)+10, \quad x_j\in[-5.12,5.12],\ j=1,\ldots,4 \tag{17}$$

The Rastrigin function is a highly multimodal and non-convex function with a large number of local minima which are regularly distributed. The function is continuous, differentiable and separable and has a global minimum value of 0. at $x_i = 0.$, $i = 1,\ldots,4$. The Rastrigin function is shown in 2D form in Figure 22.

The solution performance of both the fixed MCL and the variable MCL SA methods calculated through 100 independent computer runs are given in Table 12. Three fixed MCLs between 40 and 120 were used in this optimization experiment. It is noted that outside of this range the SA procedure was not able to reach a noticeable success rate. The variable MCL SA method was able to find the global minimum point in 13% of runs, while the best success rate for the fixed MCL SA was 9%. It is noted that finding the global minimum point of this function is rather difficult because of the large number of local minima involved in this function. However, the total average for the variable MCL SA is less than 15% of those of the fixed MCL methods. Typical convergence curves for this function are presented in Figure 23.

Table 12. The 4D Rastrigin function optimization results and statistics

MCL	40	60	120	Variable
Success rate (%)	7	6	9	13
Average function evaluations	4128.71	4891.0	6482.22	1244.08
Total average	147787	259474	208715.0	20194.1
Best values	0.000378	0.000177	0.000269	0.000279
Worst values	12.93519	8.955147	7.959662	7.959663
Mean	3.347934	2.546225	1.983397	1.843349
Variance	2.728806	1.845116	1.509851	1.522087

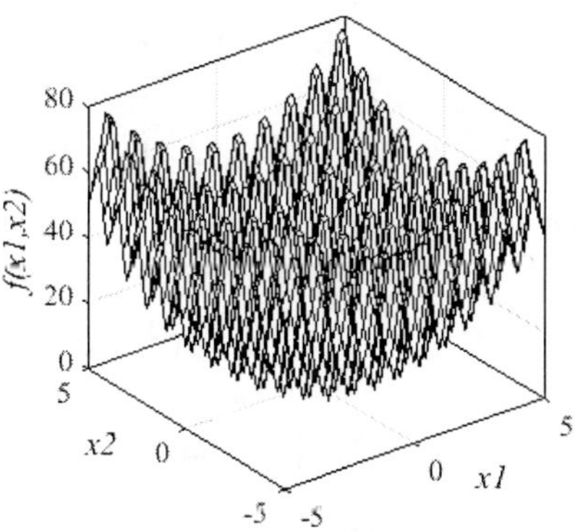

Figure 22. The 2D Rastrigin function.

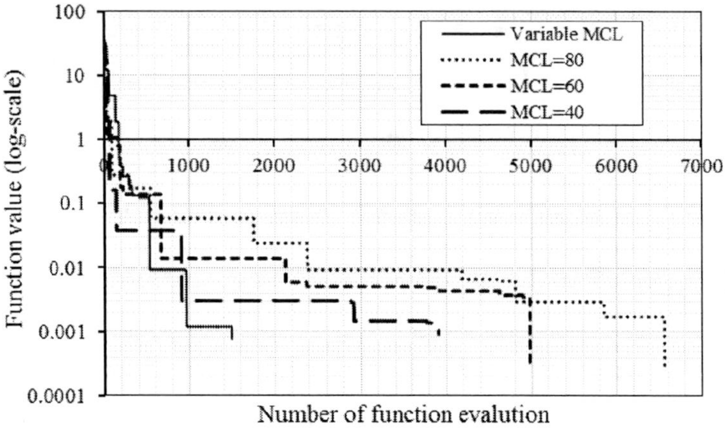

Figure 23. Typical optimization convergence curves of the 4D Rastrigin function.

11. Four-Dimensional Brown Function

$$f(x_1,...,x_4)=\sum_{j=1}^{3}[(x_j^2)^{(x_{j+1}^2+1)}+(x_{j+1}^2)^{(x_j^2+1)}],\quad x_j\in[-1,4],\quad j=1,...,4 \tag{18}$$

The Brown function is continuous, differentiable, non-separable and unimodal function [10]. The global minimum is located at $x_i = 0$, $i = 1,\ldots,4$ with the global minimum value of 0. The function is exhibited in 2D form in Figure 24.

Table 13 shows the results obtained for three different values of MCLs together with those of the variable MCL SA method. This function is very sensitive to the length of Markov chain. As the results show, for MCL = 20, the percentage of successful solution is 0., while for MCL = 80 the success rate is 100%. This increase in success rate goes together with a noticeable increase in the number of required function evaluations. The variable MCL SA method, however, presents 100% success rate with a much lower computational cost. The average function evaluations for the variable MCL SA method is only about 17% of the fixed MCL with the same success rate. Typical convergence curves for this function are plotted in Figure 25 indicating much steeper convergence to the global minimum point can be achieved by the variable MCL SA method.

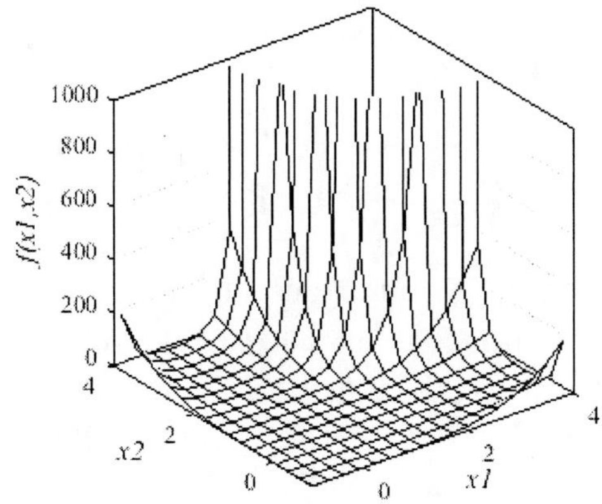

Figure 24. The 2D Brown function.

Table 13. The 4D Brown function optimization results and statistics

MCL	20	40	80	Variable
Success rate (%)	0	92	100	100
Average function evaluations	NA	4194.05	6710.66	1170.23
Total average	NA	5137.53	6710.66	1170.23
Best values	0.009663	0.000197	8.31E-05	0.000103
Worst values	1.851615	0.101971	0.000998	0.000999
Mean	0.413786	0.003494	0.000702	0.00067
Variance	0.366804	0.01373	0.000229	0.000239

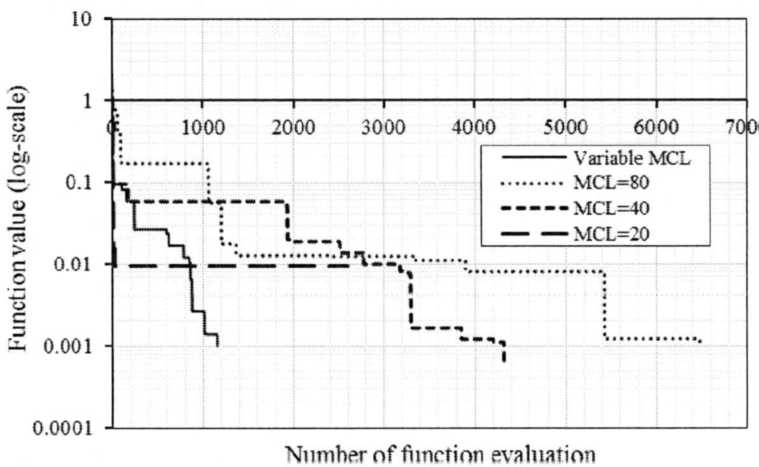

Figure 25. Typical optimization convergence curves of the Brown function.

12. Six-Dimensional Trid 6 Function

$$f(x_1,...,x_6)=\sum_{j=1}^{6}(x_j-1)^2-\sum_{j=1}^{6}x_jx_{j-1},\quad x_j\in[-36,36],\ j=1,\cdots,6 \tag{19}$$

The Trid 6 function is continuous, differentiable and non-separable function [10]. The function has only one global minimum with the value of -50. The function has been shown in 2D form in Figure 26.

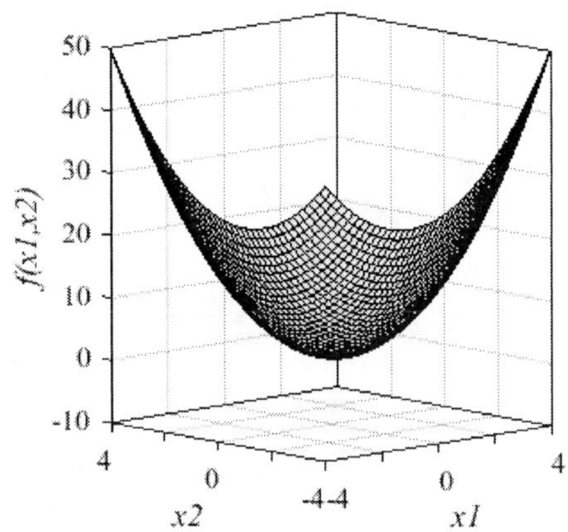

Figure 26. The Trid 6 function in 2D form.

Table 14. 6D Trid 6 function optimization results and statistics

MCL	60	90	120	Variable
Success rate (%)	24	100	100	100
Average function evaluations	9258.96	9960.51	13279.0	2322.7
Total average	60717.3	9960.51	13279.0	2322.7
Best values	-49.9992	-49.9998	-49.9999	-49.9999
Worst values	-17.4952	-49.999	-49.999	-49.999
Mean	-46.885	-49.9993	-49.9993	-49.9992
Variance	6.289948	0.000172	0.000192	0.000177

Table 14 presents the results obtained for three different values of MCL together with those of the variable MCL SA method. As the results show, this function is sensitive to the length of Markov chain. For MCL = 60, the success rate was 24%, while for MCL = 90 and 120 the success rates were 100%. The variable MCL SA method, however, presents 100% success rate with about 25% average function evaluations compred to the fixed MCL SA with the same success rate. Typical convergence curves for this function are presented in Figure 27.

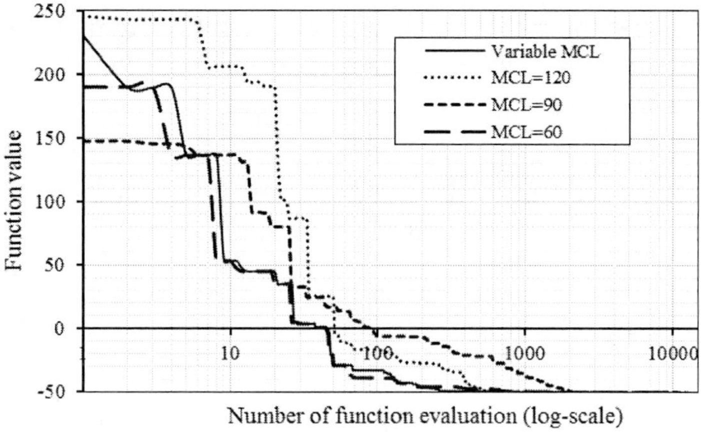

Figure 27. Typical optimization convergence curves of the Trid 6 function.

13. Six-Dimensional Chung Reynolds Function

$$f(x_1,...,x_6) = (\sum_{j=1}^{6} x_j^2)^2, \quad x_j \in [-100,100], \quad j = 1,...,6 \tag{20}$$

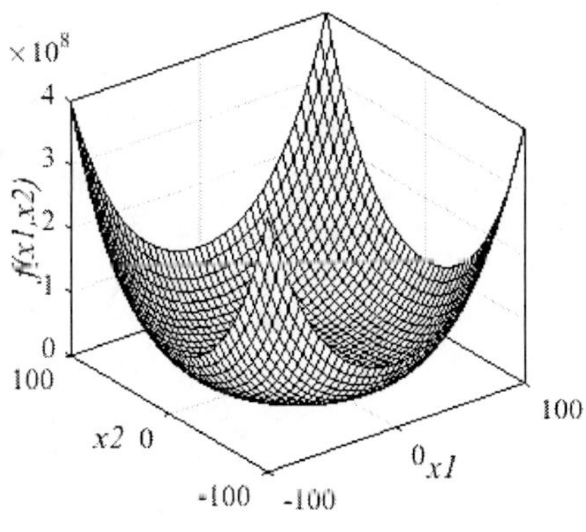

Figure 28. The Chung Reynolds function in 2D form.

The Chung Reynolds function is continuous, differentiable, partially-separable and unimodal function [10]. The global minimum is located at $x_j = 0.$, $j = 1,\ldots, 6$ with the global minimum of 0. The function is shown in 2D form in Figure 28.

The optimization results have been given in Table 15. For all three used fixed MCLs (60, 90 and 120) good success rates were obtained. The variable MCL SA strategy also reached 100% success rate but with a much lower computational cost and better solution quality. Typical convergence curves are depicted in Figure 29.

Table 15. 6D Chung Reynolds function optimization results and statistics

MCL	60	90	120	Variable
Success rate (%)	97	100	100	100
Average function evaluations	5697.45	4365.38	5738.76	1451.5
Total average	6200.03	4365.38	5738.76	1451.5
Best values	9.24E-05	3.79E-05	5.33E-05	2.51E-05
Worst values	0.006931	0.000983	0.000998	0.000996
Mean	0.000992	0.00062	0.000665	0.000574
Variance	0.000819	0.000268	0.00026	0.000272

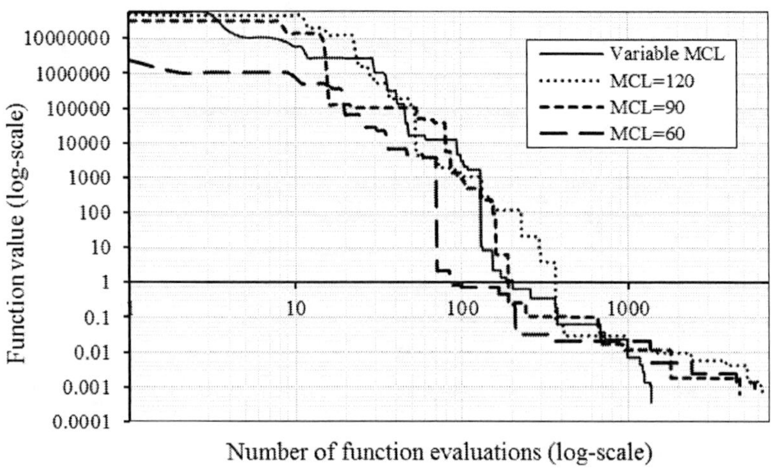

Figure 29. Typical optimization convergence curves of the 6D Chung Reynolds function.

14. Eight -Dimensional Ackley 1 Function

$$f(x_1,...,x_8) = -20\exp(-0.2\sqrt{0.125\sum_1^8 x_j^2}) - \exp(0.125\sum_1^8 \cos(2\pi x_j)) + 20 + e \tag{20}$$

The Ackley 1 function subject to $x_1,...,x_8 \in [-35,35]$ is a continuous, differentiable, non-separable and multimodal function [10]. The global minimum is located at $x_j = 0., j = 1,\ldots, 8$ with the global minimum of 0. The function has been shown in 2D form in Figure 30.

The optimization results for this function have been given in Table 16. It can be seen that the fixed MCL SA has very poor success rate for MCL = 80. However, for MCL = 96 and 120 the success rates become very good. The variable MCL SA method also showed excellent behavior with 100% success rate and about 20% function evaluations of the fixed MCL = 120. Typical convergence curves are depicted in Figure 31 indicating much better approach to the global minimum point can be achieved by using the variable MCL SA method.

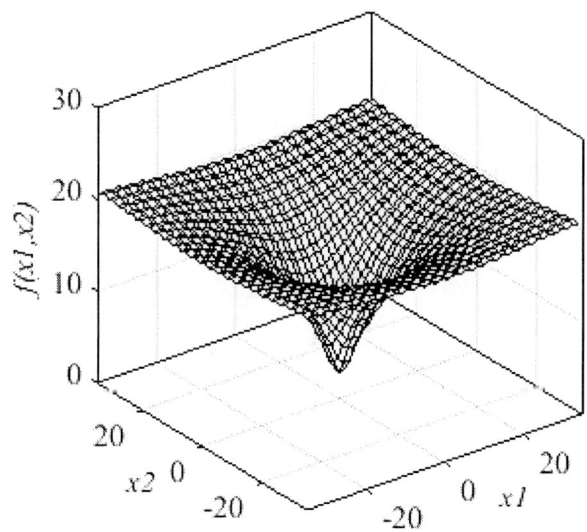

Figure 30. The Ackley 1 function in 2D form.

Table 16. 8D Ackley 1 function optimization results and statistics

MCL	80	96	120	Variable
Success rate (%)	4	88	99	100
Average function evaluations	13669.1	15037.6	18724.9	3830.39
Total average	80629.0	17765.0	18927.0	3830.39
Best values	0.000899	0.000526	0.00068	0.000626
Worst values	9.963249	10.6281	0.00114	0.000999
Mean	1.868033	0.293372	0.000914	0.000906
Variance	2.055662	1.195293	7.57E-05	7.84E-05

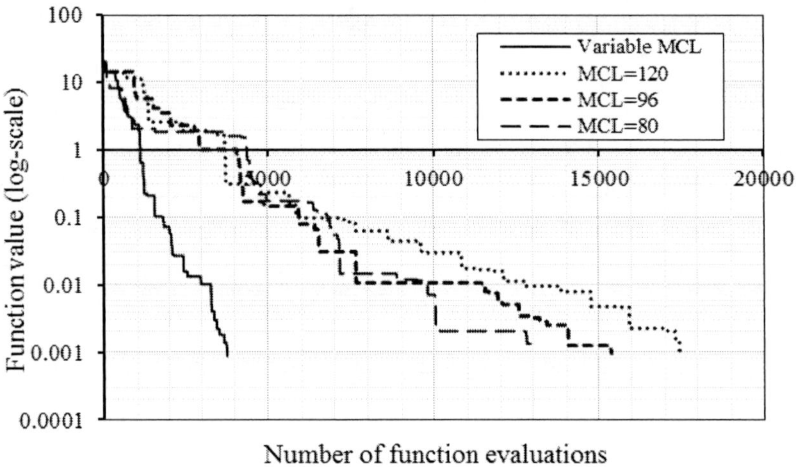

Figure 31. Typical optimization convergence curves of the 8D Ackley 1 function.

15. Eight-Dimensional Zakharov Function

$$f(x_1,\dots,x_8)=(\sum_{j=1}^{8}x_j^2)+(\sum_{j=1}^{8}0.5jx_j)^2+(\sum_{j=1}^{8}0.5jx_j)^4,\quad x_1,\dots,x_8\in[-5,10] \tag{22}$$

The Zakharov function is a continuous, differentiable and non-separable function and has one global minimum located at $x_i = 0.$, $i = 1,\dots,8$ [10]. The global minimum value is 0. The function is shown in Figure 32 in its 2D form.

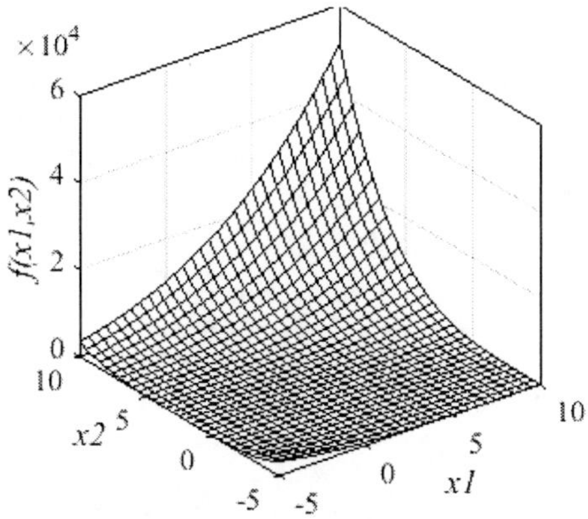

Figure 32. The Zakharov function in 2D form.

Table 17. Zakharov function optimization results and statistics

MCL	80	120	160	Variable
Success rate (%)	6	100	40	100
Average function evaluations	12766	15475.7	18947.1	4574.04
Total average	326115	15475.7	48948.6	4574.04
Best values	0.000911	0.000243	0.000309	0.00016
Worst values	45.3126	0.000998	0.002356	0.000998
Mean	5.338496	0.000841	0.001193	0.000844
Variance	8.151985	0.000146	0.000385	0.000154

Table 17 shows the results obtained for three different values of MCL together with the results of the variable MCL SA method. The fixed MCL method fails to find the global solution with a considerable success rate for MCLs less than 80. As a matter of fact, the success rate goes up with increase in the chain length and then decrease. This complex behavior reveals that the performance of the solution may have a complex relationships with the chain length which cannot be easily predicted. However, the variable MCL SA method showed excellent performance in finding the global minimum of this function. The success rate for variable method was 100% with less than 30% function evaluations as compared with the fixed MCL with equal success rate. Typical convergence curves as plotted in Figure 33 showing that very good convergence rate to the global minimum point can be attained by the variable MCL strategy.

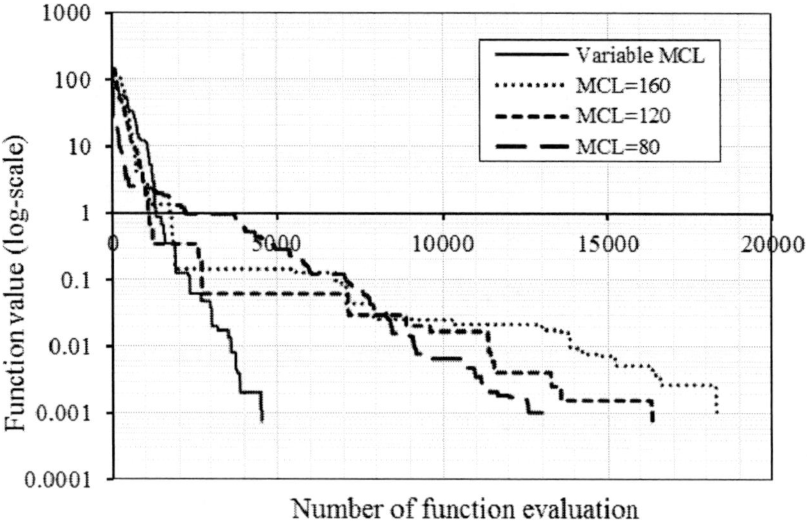

Figure 33. Typical optimization convergence curves of the 8D Zakharov function.

16. Ten-Dimensional Rosenbrock Function

In n-dimensional case the function is defined as:

$$f(x_1,\cdots,x_n)=\sum_{j=1}^{n-1}[100(x_j^2-x_{j+1})^2+(x_j-1)^2] \tag{23}$$

The Rosenbrock function is a non-convex function, introduced by Howard H. Rosenbrock [12], which is used as a performance test problem for optimization algorithms. It is also known as Rosenbrock's valley or Rosenbrock's banana function. The global minimum is inside a long, narrow, parabolic shaped flat valley. The function is defined in 2-dimensional case by $f(x_1,x_2)=(a-x_1)^2+b(x_2-x_1^2)^2$ with a minimum at $(x_1,\ x_2)=(a,\ a^2)$. The minimum function value is 0. The parameters are usually set to $a=1$ and $b=100$. The Rosenbrock function with two variables is shown in Figure 34.

The Rosenbrock function is continuous, differentiable, unimodal and non-separable function. The global minimum of function (23) is located at $x_j=1$.

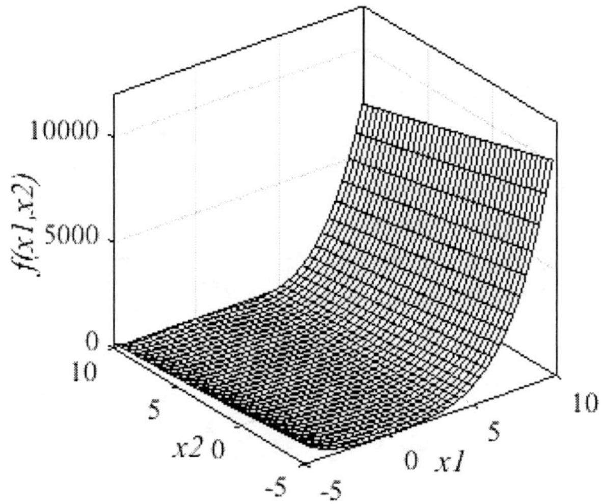

Figure 34. The Rosenbrock function in 2D form.

In this optimization case study, we used the Rosenbrock function with ten variables each raging between -5 and 10 ($x_j \in [-5,10], j = 1,\cdots,10$). The results obtained for three different values of MCL and the variable MCL SA strategy are presented in Table 18. The results show that the variable MCL SA strategy has 100% success rate with a much lower computational effort. Typical convergence curves for this function are plotted in Figure 35.

Table 18. 10D Rosenbrock function optimization results and statistics

MCL	100	130	150	Variable
Success rate (%)	48	100	26	100
Average function evaluations	16274.4	17824.6	19121.7	4505.02
Total average	37942.1	17824.6	76047.7	4505.02
Best values	0.00087	0.000568	0.000433	0.000321
Worst values	1.805325	0.000998	0.002772	0.000998
Mean	0.185324	0.000868	0.001228	0.000859
Variance	0.395004	0.000106	0.000344	0.000127

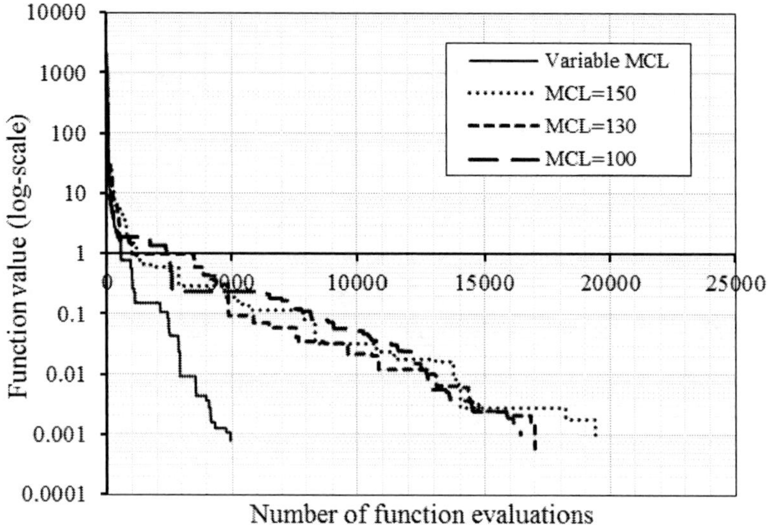

Figure 35. Typical optimization convergence curves of the Rosenbrock function.

17. Ten-Dimensional Alpine Function

$$f(x_1, x_2) = \sum_{j=1}^{n} \left| x_j \sin(x_j) + 0.1x_j \right| \tag{24}$$

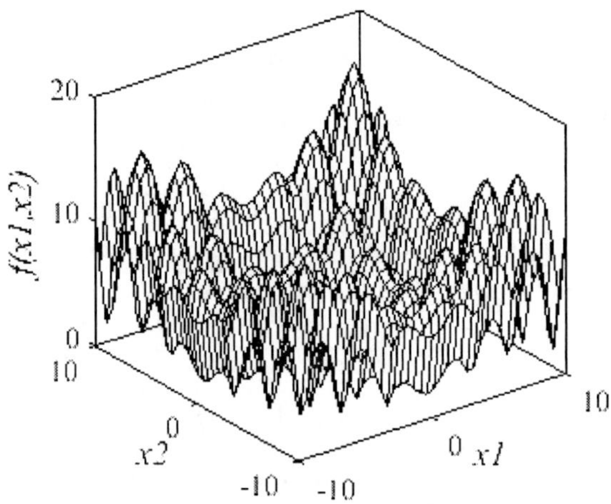

Figure 36. The Alpine function in 2D form.

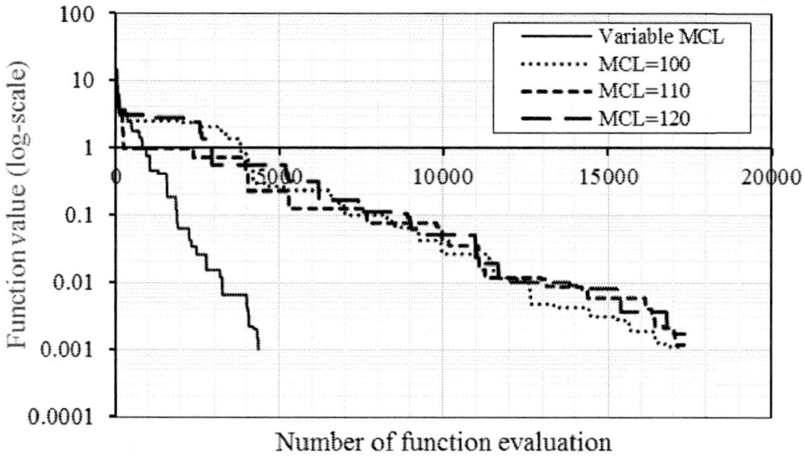

Figure 37. Typical optimization convergence curves of the 10D Alpine function.

Table 19. 10D Alpine function optimization results and statistics

MCL	100	110	120	Variable
Success rate (%)	82	96	46	100
Average function evaluations	17693.5	18594.4	19571.8	5049.72
Total average	22083.9	19427.8	43051.2	5049.72
Best values	0.000896	0.000835	0.000709	0.000729
Worst values	0.273765	0.0015	0.002577	0.001
Mean	0.008131	0.00097	0.001119	0.000931
Variance	0.033997	7.18E-05	0.000264	6.15E-05

The Alpine function can be defined on any variable number n and positive input domain. The function is usually evaluated on $x_j \in [-10,10]$, $j = 1,...,n$. The Alpine function is not convex and non-separable but differentiable. The function is multimodal and has one global minimum at $x_j = 0$. The minimum function value is 0. Figure 36 depicts the 2-dimensional form of this function.

In this case study we used Alpine function with ten variables. Again, three fixed MCLs were used and the success rates and other performance parameters were determined. The best interval for fixed MCLs for optimization of this function was found to be between 100 and 110. In this interval, the best calculated success rate was 96% with the total average of

19427.8. However, the success rate for the variable MCL SA method was 100% with total average of 5049.7. The all results are given in Table 19. Figure 37 shows typical convergence curves showing that the convergence to the global optimum point is much faster than the other methods.

18. Sixteen-Dimensional Csendes Function

$$f(x_1,\ldots,x_{16}) = \sum_{j=1}^{16} x_j^6 \left(2+\sin\frac{1}{x_j}\right), \quad x_j \in [-1,1], \ j=1,\ldots,16 \tag{25}$$

The Csendes function is a continuous, differentiable, separable and multimodal function [10]. This function is usually evaluated on $x_j \in [-1,1],\ j=1,\ldots,16$. It has one global minimum at $x_j = 0$, and its minimum function value is 0. Figure 38 depicts the 2D form of this function.

This function is found to be very sensitive to the value of MCLs. After extensive trial and error calculations, three fixed MCLs have been found such that some successful solutions were reached. The best success rate for fixed MCL SA method was 21% with total average 93087.4. The variable MCL SA strategy however presents 100% success rate with total average of 4491.05, which is only about 5% of that of the MCL-196. The all results are given in Table 20. Figure 37 also shows typical convergence curves showing that the convergence to the global optimum point by using the variable MCL SA method is much faster than the other methods.

Table 20. 16D Csendes function optimization results and statistics

MCL	160	176	196	Variable
Success rate (%)	4	17	21	100
Average function evaluations	18947.2	18112.7	17845.6	4491.05
Total average	498971	115765	93087.4	4491.05
Best values	0.000874	0.00063	0.000568	0.000201
Worst values	0.028949	0.014886	0.007419	0.000999
Mean	0.006773	0.002667	0.00227	0.000815
Variance	0.00564	0.002233	0.001515	0.000186

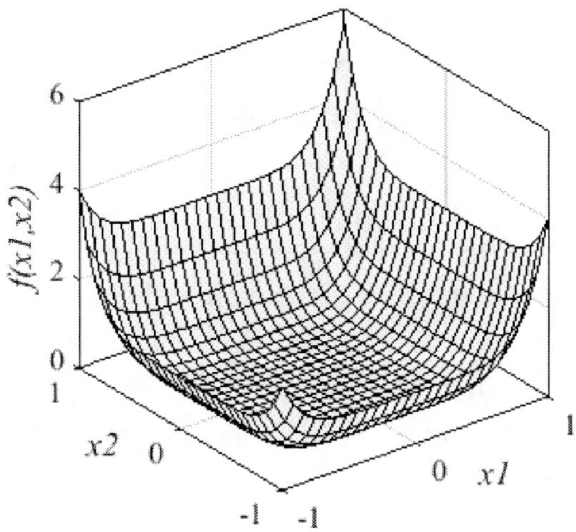

Figure 38. The Csendes function in its 2D form.

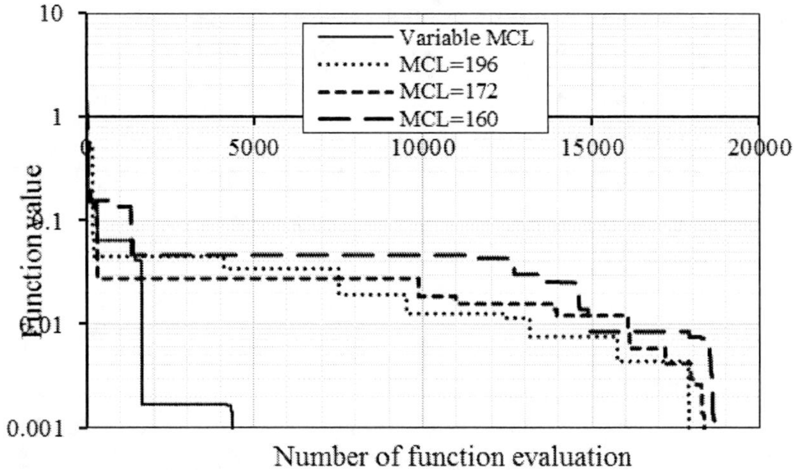

Figure 39. Typical optimization convergence curves of the 16D Csendes function.

Discussion

In SA optimization method, most features such as state space, candidate points and cost function are fixed by definition. The only feature which is variable during the calculation is the temperature. Therefore, one of the most important tips in SA is the adaptation of the cooling strategy, and many attempts have been made to develop or suggest good cooling strategies. The most popular annealing procedure involves first applying relatively high temperature to the system and then lowering the temperature sequentially by a constant factor R ($0<R<1$) (see Eq. 2), taking enough steps at each temperature to keep the system close to equilibrium, until the system approaches to the ground state. It is believed that equilibrium can be achieved by keeping the temperature constant for a limited number of iterations, say 20-80 iterations. The first issue that must be identified is how sensitive the SA can be to the length of these iterations. The results of the considered benchmark functions show that by increasing the chain length the success rates and total averages are generally increased. Among them, increase in the total averages can be very severe.

It is generally stated that the success rate is a very important parameter to show the power of an optimization algorithm. The obtained results reveal that in most of the benchmark functions the variable MCL SA method improves the success rate of the solution. For example, for 16D Csendes function the best success rate found by the fixed MCL SA was 21%, while this percentage was 100 in the variable MCL SA method. Similarly, the best success rates found for 4D Rastrigin function by the fixed MCL SA and the proposed variable MCL SA were 9% and 13%, respectively. It is noted that the 4D Rasrigin function possess a large number of local minima and is generally known as a difficult problem for many of the optimization algorithms.

The results show that the average function evaluations can significantly be decreased by using the variable MCL SA method. In most cases, the average of function evaluations decreased by a factor between 2 and 4. Similarly, the total average, in which the failed runs have also been taken into account, showed significant reduction by using the variable MCL SA method. For example, total averages for optimization of 16D Csendes and 4D Rastgirin functions decreased more than 90% and 86%, respectively, as compared to the fixed MCL SA method. For other benchmark functions, various degrees of reduction in total averages have been achieved.

The other performance factor is the solution quality which can be measured by the best and the worst solutions and their statistical parameters such as mean and variance. The results obtained in this study showed that by using the variable MCL SA method the solution qualities have been improved for many test functions such as 16D Csendes, 10D Alpine and 10D Rosenbrock. However, for some test functions such as 2D W/Wavy and 2D Bohachevsky functions, this is not the case. Overall, it can be stated that the solution quality of the proposed variable MCL SA is comparable with that of the fixed MCL SA method.

CONCLUSION

In this study we aimed at evaluating a variable Markov chain length SA method for unconstrained optimization problems. For this purpose, we applied the variable MCL SA method to eighteen benchmark functions with different dimensions. The optimization results and solution performance parameters have been compared to the fixed MCL SA method with three different chain lengths. The experimental results show that the variable MCL SA method introduces a considerable improvement to the SA method in terms of both success rate and computational cost. The proposed variable MCL SA method, in fact, eliminates an experimental parameter of SA procedure and makes it easier to implement. Overall, we believe that the variable MCL SA method is a step forward to acquire an effective and reliable method for unconstrained optimization purposes.

REFERENCES

[1] Batholomew-Biggs, M., (2009). *Nonlinear optimization with engineering application.* Springer US.

[2] Arora J. S., (2017). Chapter 16-Global optimization concepts and methods. In: *Introduction to optimum design* (fourth edition, pp. 707-738). Academic Press.

[3] Rassias, T., Floudas, C. A., Butenko, S., (2014). *Optimization in science and engineering.* Springer US.

[4] Kirkpatrick, S., Gallat, C. D., Vecchi, M. P., (1983). Optimization by simulated annealing. *Science.* 220, 671-680.

[5] Javidrad F., Nouri R. (2011). A simulated annealing method for design of laminated with required stiffness properties. *Compos. Struct.* 93, 1127-1135.

[6] Hajek B. (1988). Cooling schedules for optimal annealing. *Math. Operat. Res.* 13, 311-329.

[7] Ingber L., (1989). Very fast simulated re-annealing. *Mathl. Comput. Modelling.* 12(8), 967-973.

[8] Azizi N., Zolfaghari S., (2004). Variable temperature control for simulated annealing: a comparative study. *Compu. Operat. Res.* 31, 2439-2451.

[9] Lee D., Wexler A. S., (2011). Simulated annealing implementation with shorter MCL to reduce computational burden and its application to the analysis of pulmonary airway architecture. *Compu. Bio. Med.* 41, 707-715.

[10] Jamil M., Yang X. S., (2013). A Literature Survey of Benchmark Functions for Global Optimization Problems. *Int. J. of Math. Modelling and Num. Optim.* 4(2), 150–194.

[11] Styblinski, M. A., Tang, T. S., (1990). Experiments in nonconvex optimization: Stochastic approximation with function smoothing and simulated annealing, *Neural Networks* 3(4), 467-483.

[12] Rosenbrock, H. H., (1960). An automatic method for finding the greatest or least value of a function. *The Comput. J.* 196, 3(3), 175–184.

INDEX

A

B

C

D

E

F

G

H

I

L

M

N

O

P

Q

R

S

T

U

V